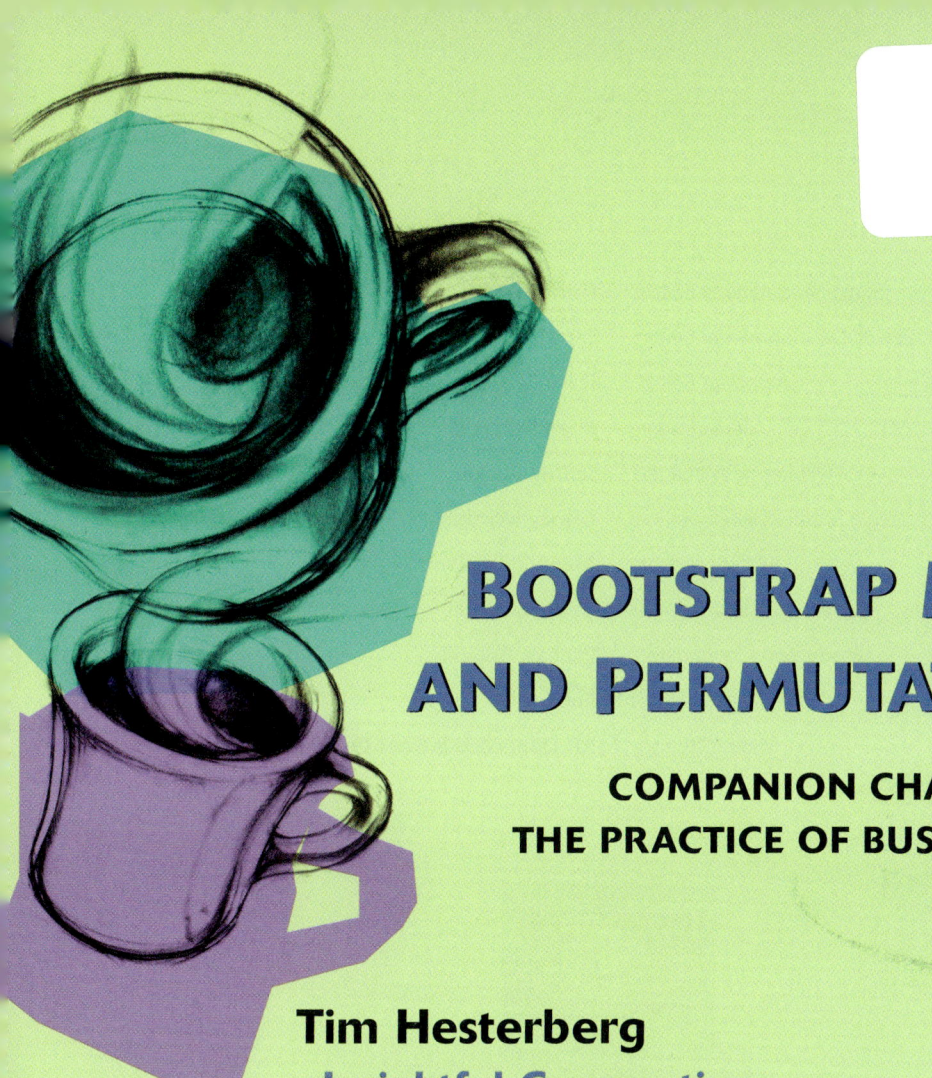

BOOTSTRAP METHODS AND PERMUTATION TESTS

COMPANION CHAPTER 18 TO THE PRACTICE OF BUSINESS STATISTICS

Tim Hesterberg
Insightful Corporation

Shaun Monaghan
Eastlake High School

David S. Moore
Purdue University

Ashley Clipson
University of Puget Sound

Rachel Epstein
Reed College

W. H. Freeman and Company
New York

Senior Acquisitions Editor:	**Patrick Farace**
Senior Developmental Editor:	**Terri Ward**
Associate Editor:	**Danielle Swearengin**
Media Editor:	**Brian Donnellan**
Marketing Manager:	**Jeffrey Rucker**
Head of Strategic Market Development:	**Clancy Marshall**
Project Editor:	**Mary Louise Byrd**
Cover and Text Design:	**Vicki Tomaselli**
Production Coordinator:	**Paul W. Rohloff**
Composition:	**Publication Services**
Manufacturing:	**RR Donnelley & Sons Company**

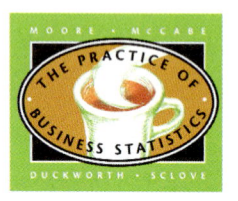

S-PLUS™ is a registered trademark of the Insightful Corporation.

Cataloguing-in-Publication Data available from the Library of Congress

Library of Congress Control Number: 2002108463

©2003 by Tim Hesterberg. All rights reserved.

This chapter was written by Tim Hesterberg, Shaun Monaghan, David S. Moore, Ashley Clipson, and Rachel Epstein, with support from the National Science Foundation under grant DMI-0078706. We thank Bob Thurman, Richard Heiberger, Laura Chihara, Tom Moore, and Gudmund Iversen for helpful comments.

No part of this book may be reproduced by any mechanical, photographic, or electronic process, or in the form of a phonographic recording, nor may it be stored in a retrieval system, transmitted, or otherwise copied for public or private use, without the written permission of the publisher.

Printed in the United States of America

First Printing 2003

TO THE INSTRUCTOR

NOW *YOU* HAVE THE CHOICE!

This is **Companion Chapter 18** to *The Practice of Business Statistics (PBS)*. Please note that this chapter, along with any other Companion Chapters, can be bundled with the *PBS* Core book, which contains Chapters 1–11.

CORE BOOK
- Chapter 1   Examining Distributions
- Chapter 2   Examining Relationships
- Chapter 3   Producing Data
- Chapter 4   Probability and Sampling Distributions
- Chapter 5   Probability Theory
- Chapter 6   Introduction to Inference
- Chapter 7   Inference for Distributions
- Chapter 8   Inference for Proportions
- Chapter 9   Inference for Two-Way Tables
- Chapter 10   Inference for Regression
- Chapter 11   Multiple Regression

These other **Companion Chapters**, *in any combinations you wish,* are available for you to package with the *PBS* Core book.

TAKE YOUR PICK
- Chapter 12   Statistics for Quality: Control and Capability
- Chapter 13   Time Series Forecasting
- Chapter 14   One-Way Analysis of Variance
- Chapter 15   Two-Way Analysis of Variance
- Chapter 16   Nonparametric Tests
- Chapter 17   Logistic Regression
- Chapter 18   Bootstrap Methods and Permutation Tests

BOOTSTRAP METHODS AND PERMUTATION TESTS

18.1 Why Resampling? 18-4
    Note on software 18-5

18.2 Introduction to Bootstrapping 18-5
    Case 18.1 Telecommunication Repair Times 18-5
    Procedure for bootstrapping 18-7
    Using software 18-10
    Why does bootstrapping work? 18-11
    Sampling distribution and bootstrap distribution 18-12
        Section 18.2 Exercises 18-14

18.3 Bootstrap Distributions and Standard Errors 18-16
    Case 18.2 Real Estate Sale Prices 18-17
    Bootstrap distributions of other statistics 18-20
    Bootstrap t confidence intervals 18-22
    Bootstrapping to compare two groups 18-23
    Beyond the basics: the bootstrap for a scatterplot smoother 18-27
        Section 18.3 Exercises 18-29

18.4 How Accurate Is a Bootstrap Distribution? 18-32
    Bootstrapping small samples 18-34
    Bootstrapping a sample median 18-36
        Section 18.4 Exercises 18-38

18.5 Bootstrap Confidence Intervals 18-39
    Bootstrap percentiles as a check 18-39
    Confidence intervals for the correlation coefficient 18-41
    Case 18.3 Baseball Salaries and Performance 18-41
    More accurate bootstrap confidence intervals 18-44
    Bootstrap tilting and BCa intervals 18-45
    How the BCa and tilting intervals work 18-47
        Section 18.5 Exercises 18-48

18.6 Significance Testing Using Permutation Tests 18-51
    Using software 18-55
    Permutation tests in practice 18-57
    Permutation tests in other settings 18-61
        Section 18.6 Exercises 18-65

Statistics in Summary 18-68

Chapter 18 Review Exercises 18-68

Notes for Chapter 18 18-73

Solutions to Odd-Numbered Exercises S-18-1

Prelude

Ensuring equal service

In 1984, in an effort to open the telecommunications market to competition, AT&T was split into eight regional companies. To promote competition, more than one company was now allowed to offer telecommunications services in a local market. But since it isn't in the public interest to have multiple companies digging up local streets to bury cables, one telephone company in each region is given responsibility for installing and maintaining all local lines and leasing capacity to other carriers.

Each state's Public Utilities Commission (PUC) is responsible for seeing that there is fair access for all carriers. For example, the primary carrier should do repairs as quickly for customers of other carriers as for their own. Significance tests are used to compare the levels of service. If a test indicates that service levels are not equivalent, the primary carrier pays a penalty.

PUCs and primary carriers perform many of these tests each day. Given the large amounts of money at stake, the significance tests described in earlier chapters are not sufficiently accurate. Instead, primary carriers like Verizon have turned to resampling methods in an effort to achieve accurate test results that provide a strong defense in an adversarial hearing before a PUC.

The resampling methods of this chapter provide alternatives to the methods of earlier chapters for finding standard errors and confidence intervals and for performing significance tests.

# CHAPTER 18

Bootstrap Methods and Permutation Tests

- 18.1 Why Resampling?
- 18.2 Introduction to Bootstrapping
- 18.3 Bootstrap Distributions and Standard Errors
- 18.4 How Accurate Is a Bootstrap Distribution?
- 18.5 Bootstrap Confidence Intervals
- 18.6 Significance Testing Using Permutation Tests

18.1 Why Resampling?

Statistics is changing. Modern computers and software make it possible to look at data graphically and numerically in ways previously inconceivable. They let us do more realistic, accurate, and informative analyses than can be done with pencil and paper.

The bootstrap, permutation tests, and other resampling methods are part of this revolution. Resampling methods allow us to quantify uncertainty by calculating standard errors and confidence intervals and performing significance tests. They require fewer assumptions than traditional methods and generally give more accurate answers (sometimes very much more accurate). Moreover, resampling lets us tackle new inference settings easily. For example, Chapter 7 presented methods for inference about the difference between two population means. But suppose you are really interested in a *ratio* of means, such as the ratio of average men's salary to average women's salary. There is no simple traditional method for inference in this new setting. Resampling not only works, but works in the same way as for the difference in means. We don't need to learn new formulas for every new problem.

Resampling also helps us understand the concepts of statistical inference. The sampling distribution is an abstract idea. The bootstrap analog (the "bootstrap distribution") is a concrete set of numbers that we analyze using familiar tools like histograms. The standard deviation of that distribution is a concrete analog to the abstract concept of a standard error. Resampling methods for significance tests have the same advantage; permutation tests produce a concrete set of numbers whose "permutation distribution" approximates the sampling distribution under the null hypothesis. Comparing our statistic to these numbers helps us understand *P*-values. Here is a summary of the advantages of these new methods:

- **Fewer assumptions.** For example, resampling methods do not require that distributions be Normal or that sample sizes be large.

- **Greater accuracy.** Permutation tests, and some bootstrap methods, are more accurate in practice than classical methods.

- **Generality.** Resampling methods are remarkably similar for a wide range of statistics and do not require new formulas for every statistic. You do not need to memorize or look up special formulas for each procedure.

- **Promote understanding.** Bootstrap procedures build intuition by providing concrete analogies to theoretical concepts.

Resampling has revolutionized the range of problems accessible to business people, statisticians, and students. It is beginning to revolutionize our standards of what is acceptable accuracy in high-stakes situations such as legal cases, business decisions, and clinical trials.

Note on software

Bootstrapping and permutation tests are feasible only with the use of software to automate the heavy computation that these resampling methods require. If you are sufficiently expert in programming or with a spreadsheet, you can program basic resampling methods yourself. But it is easier to use software with resampling methods built in.

This chapter uses S-PLUS,[1] the software choice of most statisticians doing research on resampling methods. A free student version of this software is available to students and faculty at `elms03.e-academy.com/splus/`. In addition, a student library containing data sets specifically for your book, menu-driven access to capabilities you'll need, and a manual that accompanies this chapter can be found at `www.insightful.com/Hesterberg/bootstrap` or at `www.whfreeman.com`. You may also order an S-PLUS manual[2] to supplement this book from `www.whfreeman.com`.

18.2 Introduction to Bootstrapping

Let's get a feel for bootstrapping by seeing how it works in a specific example. We'll begin by showing how to bootstrap and then relate the results to ideas you've already encountered, such as standard errors and sampling distributions.

CASE 18.1 TELECOMMUNICATION REPAIR TIMES

Verizon is the primary local telephone company (the legal term is Incumbent Local Exchange Carrier, ILEC) for a large area in the eastern United States. As such, it is responsible for providing repair service for the customers of other telephone companies (known as Competing Local Exchange Carriers, CLECs) in this region. Verizon is subject to fines if the repair times (the time it takes to fix a problem) for CLEC customers are substantially worse than those for Verizon's own customers. This is determined using hypothesis tests, negotiated with the local Public Utilities Commission (PUC).

We begin our analysis by focusing on Verizon's own customers. Figure 18.1 shows the distribution of a random sample of 1664 repair times.[3] The data file is *verizon.dat*. A quick glance at the distribution reveals that the data are far from Normal. The distribution has a long right tail (skewness to the right).

The mean repair time for Verizon customers in this sample is $\bar{x} = 8.41$ hours. This is a statistic from just one random sample (albeit a fairly large one). The statistic $\bar{x}$ will vary if we take more samples, and its trustworthiness as an estimator of the population mean μ depends on how much it varies from sample to sample.

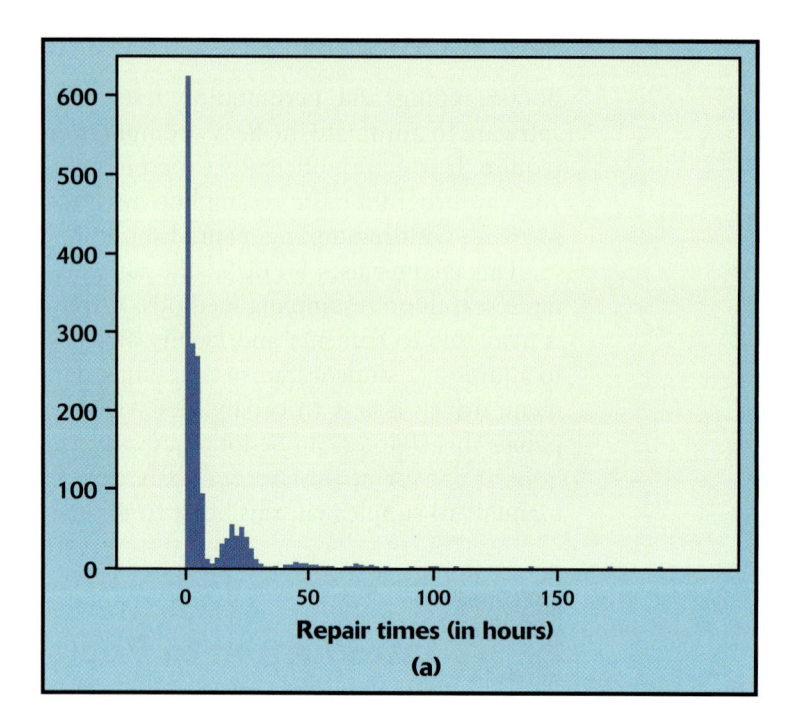

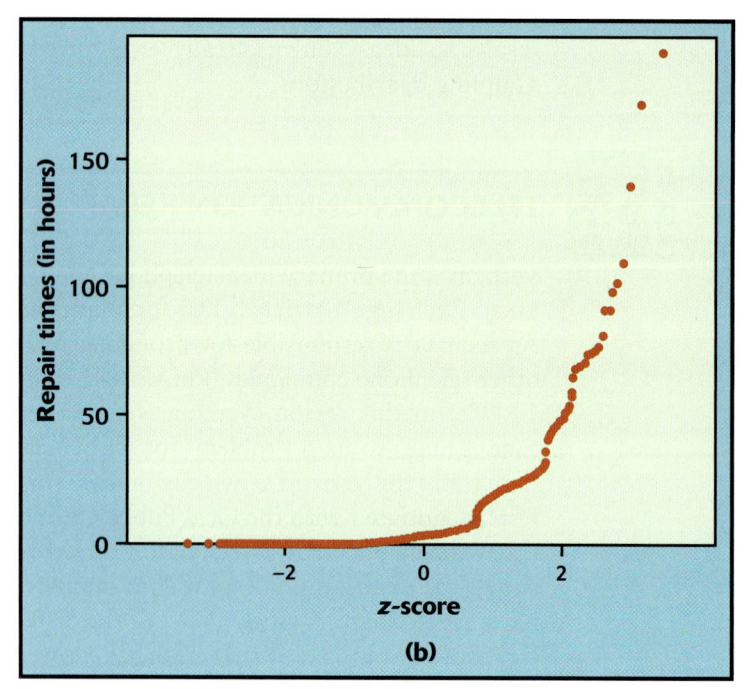

FIGURE 18.1 (a) The distribution of 1664 repair times for Verizon customers. (b) Normal quantile plot of the repair times. The distribution is clearly right-skewed rather than Normal.

Procedure for bootstrapping

Statistical inference is based on the sampling distributions of sample statistics. The bootstrap is first of all a way of finding the sampling distribution, at least approximately, from just one sample. Here is the procedure:

resamples

Step 1: Resample. Create hundreds of new samples, called **bootstrap samples** or **resamples**, by sampling *with replacement* from the original random sample. Each resample is the same size as the original random sample.

sampling with replacement

Sampling with replacement means that after we randomly draw an observation from the original sample, we put it back before drawing the next observation. This is like drawing a number from a hat, then putting it back before drawing again. As a result, any number can be drawn once, more than once, or not at all. If we sampled *without* replacement, we'd get the same set of numbers we started with, though in a different order. Figure 18.2 illustrates the bootstrap resampling process on a small scale. In practice, we would start with the entire original sample, not just six observations, and draw hundreds of resamples, not just three.

Step 2: Calculate the bootstrap distribution. Calculate the statistic for each resample. The distribution of these resample statistics is called a **bootstrap distribution**. In Case 18.1, we want to estimate the population mean repair time μ, so the statistic is the sample mean $\bar{x}$.

bootstrap distribution

Step 3: Use the bootstrap distribution. The bootstrap distribution gives information about the shape, center, and spread of the sampling distribution of the statistic.

THE BOOTSTRAP IDEA

The original sample represents the population from which it was drawn. So resamples from this sample represent what we would get if we took many samples from the population. The bootstrap distribution of a statistic, based on many resamples, represents the sampling distribution of the statistic, based on many samples.

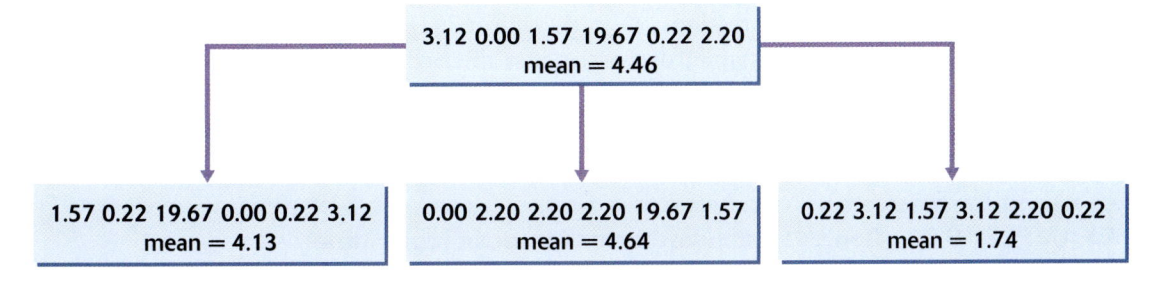

3.12 0.00 1.57 19.67 0.22 2.20
mean = 4.46

1.57 0.22 19.67 0.00 0.22 3.12
mean = 4.13

0.00 2.20 2.20 2.20 19.67 1.57
mean = 4.64

0.22 3.12 1.57 3.12 2.20 0.22
mean = 1.74

FIGURE 18.2 The resampling idea. The top box is a sample of size $n = 6$ from the Verizon data. The three lower boxes are three resamples from this original sample. Some values from the original sample occur more than once in the resamples because each resample is formed by sampling with replacement. We calculate the statistic of interest—the sample mean in this example—for the original sample and each resample.

EXAMPLE 18.1

CASE 18.1

Bootstrap distribution for mean repair time

Figure 18.3 displays the bootstrap distribution of 1000 resample means for the Verizon repair time data, using a histogram and a density curve on the top and a Normal quantile plot on the bottom. The solid vertical line in the top panel marks the mean of the original sample, and the dashed line marks the mean of the bootstrap means.

Shape: We see that the bootstrap distribution is nearly Normal. The central limit theorem says that the sampling distribution of the sample mean $\bar{x}$ is approximately Normal if n is large. So the bootstrap distribution shape is close to the shape we expect the sampling distribution to have.

Center: The bootstrap distribution is centered close to the mean of the original sample. That is, the mean of the bootstrap distribution has little bias as an estimator of the mean of the original sample. We know that the sampling distribution of $\bar{x}$ is centered at the population mean μ, that is, that $\bar{x}$ is an unbiased estimate of μ. So the resampling distribution again behaves (starting from the original sample) as we expect the sampling distribution to behave (starting from the population).

Spread: Figure 18.3 gives a rough idea of the variation among the resample means. We can get a more precise idea by computing the standard deviation of the bootstrap distribution. Applying the bootstrap idea, we use this standard deviation to estimate the standard deviation of the sampling distribution of $\bar{x}$.

> **BOOTSTRAP STANDARD ERROR**
>
> The **bootstrap standard error** of a statistic is the standard deviation of the bootstrap distribution of that statistic.

If the statistic of interest is the sample mean $\bar{x}$, the bootstrap standard error based on B resamples is

$$\text{SE}_{\text{boot},\bar{x}} = \sqrt{\frac{1}{B-1}\sum\left(\bar{x}^* - \frac{1}{B}\sum \bar{x}^*\right)^2}$$

In this expression, $\bar{x}^*$ is the mean value of an individual resample. The bootstrap standard error is just the ordinary standard deviation of the B values of $\bar{x}^*$. The asterisk in $\bar{x}^*$ distinguishes the mean of a resample from the mean $\bar{x}$ of the original sample.

EXAMPLE 18.2

CASE 18.1

Bootstrap standard error for mean repair time

The bootstrap standard error for the 1000 resample means displayed in Figure 18.3 case is $\text{SE}_{\text{boot},\bar{x}} = 0.367$. This estimates the standard deviation of the sampling distribution of $\bar{x}$.

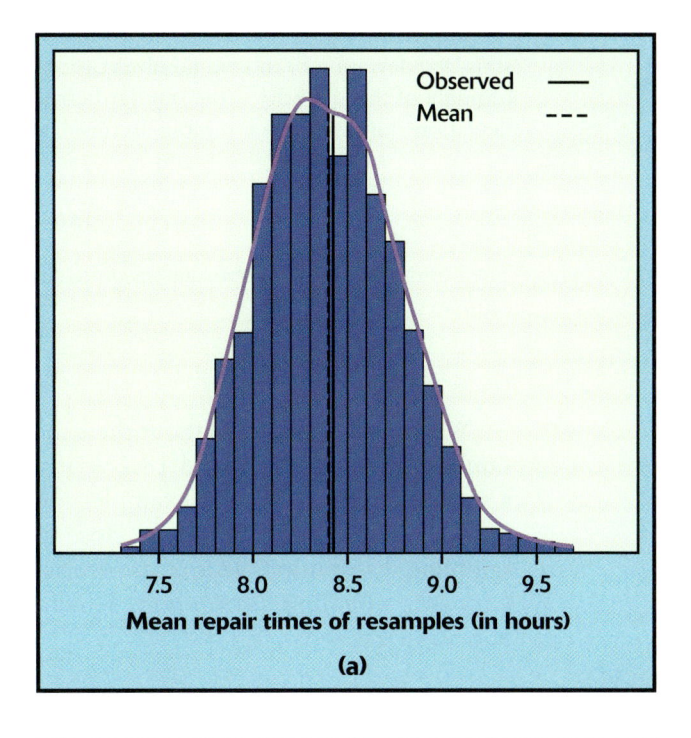

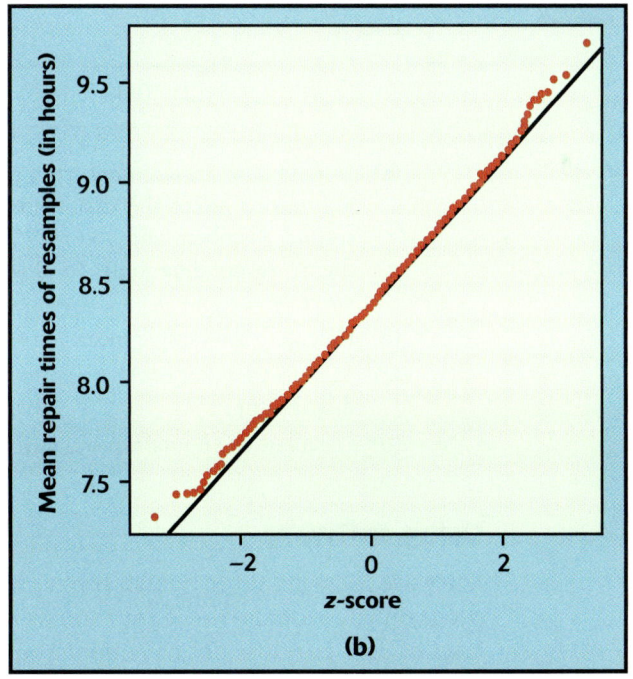

FIGURE 18.3 The bootstrap distribution for 1000 resample means from the Verizon ILEC sample. The solid line in the top panel marks the original sample mean, and the dashed line marks the average of the bootstrap means. The Normal quantile plot confirms that the bootstrap distribution is nearly Normal in shape.

In fact, we know that the standard deviation of $\bar{x}$ is $\sigma/\sqrt{n}$, where σ is the standard deviation of individual observations in the population. Our usual estimate of this quantity is the standard error of $\bar{x}$, $s/\sqrt{n}$, based on the standard deviation s of the original sample. In this example,

$$\frac{s}{\sqrt{n}} = \frac{14.69}{\sqrt{1664}} = 0.360$$

The bootstrap standard error agrees quite closely with this formula-based estimate.

We know a great deal about the behavior of the sample mean $\bar{x}$ in large samples. Examples 18.1 and 18.2 verify the bootstrap idea for the mean of a sample of size 1664. The examples show that the shape, bias, and spread of the bootstrap distribution are close to the shape, bias, and spread of the sampling distribution. *This is also true in many situations where we do not know the sampling distribution.* This fact is the basis of the usefulness of bootstrap methods.

APPLY YOUR KNOWLEDGE

18.1 Bootstrap a small data set by hand. To illustrate the bootstrap procedure, let's bootstrap a small random subset of the Verizon data:

| 3.12 | 0.00 | 1.57 | 19.67 | 0.22 | 2.20 |

(a) Sample *with replacement* from this initial SRS by rolling a die. Rolling a 1 means select the first member of the SRS, a 2 means select the second member, and so on. (You can also use Table B of random digits, responding only to digits 1 to 6.) Create 20 resamples of size $n = 6$.
(b) Calculate the sample mean for each of the resamples.
(c) Make a stemplot of the means of the 20 resamples. This is the bootstrap distribution.
(d) Calculate the bootstrap standard error.

Using software

Software is essential for bootstrapping in practice. Here is an outline of the program you would write if your software will choose random samples from a set of data but does not have bootstrap functions:

```
Repeat 1000 times {
   Draw a resample with replacement from the data.
   Calculate the resample mean.
   Save the resample mean into a vector (a variable).
}
Make a histogram and Normal quantile plot of the 1000 means.
Calculate the standard deviation of the 1000 means.
```

```
Number of Replications: 1000

Summary Statistics:
          Observed    Mean      Bias       SE
mean         8.412    8.395   -0.01698   0.3672


Percentiles:
          2.5%   5.0%   95.0%   97.5%
mean     7.717  7.814   9.028   9.114
```

FIGURE 18.4 S-PLUS output for the Verizon data bootstrap, Case 18.1.

EXAMPLE 18.3 **Using S-PLUS**

CASE 18.1

Suppose that we save the 1664 Verizon repair times as the variable ILEC in S-PLUS (commands to do this are in the file *readdata.ssc*). We can make 1000 resamples and analyze their means using these commands:

```
bootILEC = bootstrap(data = ILEC, statistic = mean)
plot(bootILEC)
qqnorm(bootILEC)
summary(bootILEC)
```

The same functions are available in menus, but it is a bit easier to discuss the typed commands. The first command resamples from the ILEC data set, calculates the means of the resamples, and saves the bootstrap results as the object named `bootILEC`. By default, S-PLUS takes 1000 resamples. The remaining three commands make a histogram (with a density curve) and a Normal quantile plot and calculate numerical summaries. The summaries include the bootstrap standard error.

Figure 18.4 is part of the output of the `summary` command. The `Observed` column gives the mean $\bar{x} = 8.412$ of the original sample. `Mean` is the mean of the resample means. The `Bias` column shows the difference between the `Mean` and the `Observed` values. The bootstrap standard error is displayed in the `SE` column. The `Percentiles` are percentiles of the bootstrap distribution, that is, of the means of the 1000 resamples pictured in Figure 18.3. All of these values except `Observed` will differ a bit if you repeat 1000 resamples, because resamples are drawn at random.

APPLY YOUR KNOWLEDGE

18.2 **Earnings for white female hourly workers.** Bootstrap the mean of the white female hourly workers data from Table 1.8 (page 31).

CASE 1.2

(a) Plot the bootstrap distribution (histogram or density plot and Normal quantile plot). Is it approximately Normal?
(b) Find the bootstrap standard error.
(c) Find the 2.5th and 97.5th percentiles of the bootstrap distribution.

Why does bootstrapping work?

It might seem that the bootstrap creates data out of nothing. This seems suspicious. But we are not using the resampled observations as if they were real data—the bootstrap is not a substitute for gathering more data to improve accuracy. Instead, the bootstrap idea is to use the resample means to estimate how the sample mean of a sample of size 1664 from this population varies because of random sampling.

Using the data twice—once to estimate the population mean, and again to estimate the variation in the sample mean—is perfectly legitimate. Indeed, we've done this many times before: for example, when we calculated both $\bar{x}$ and $s/\sqrt{n}$ from the same data. What is different is that

1. we compute a standard error by using resampling rather than the formula $s/\sqrt{n}$, and

2. we use the bootstrap distribution to see whether the sampling distribution is approximately Normal, rather than just hoping that our sample is large enough for the central limit theorem to apply.

The bootstrap idea applies to statistics other than sample means. To use the bootstrap more generally, we appeal to another principle—one that we have often applied without thinking about it.

> **THE PLUG-IN PRINCIPLE**
>
> To estimate a parameter, a quantity that describes the population, use the statistic that is the corresponding quantity for the sample.

The plug-in principle suggests that we estimate a population mean μ by the sample mean $\bar{x}$ and a population standard deviation σ by the sample standard deviation s. Estimate a population median by the sample median. To estimate the standard deviation of the sample mean for an SRS, $\sigma/\sqrt{n}$, plug in s to get $s/\sqrt{n}$. The bootstrap idea itself is a form of the plug-in principle: substitute the distribution of the data for the population distribution, then draw samples (resamples) to mimic the process of building a sampling distribution. Let's look at this more closely.

Sampling distribution and bootstrap distribution

Confidence intervals, hypothesis tests, and standard errors are all based on the idea of the *sampling distribution* of a statistic—the distribution of values taken by the statistic in all possible samples of the same size from the same population. Figure 18.5(a) shows the idea of the sampling distribution of the sample mean $\bar{x}$. In practice, we can't take a large number of random samples in order to construct this sampling distribution. Instead, we have used a shortcut: if we start with a model for the distribution of the population, the laws of probability tell us (in some situations) what the sampling distribution is. Figure 18.5(b) illustrates an important situation in which this approach works. If the population has a Normal distribution, then the sampling distribution of $\bar{x}$ is also Normal.

In many settings, we have no model for the population. We then can't appeal to probability theory, and we also can't afford to actually take many samples. The bootstrap rescues us. Use the one sample we have as though it were the population, taking many resamples from it to construct the

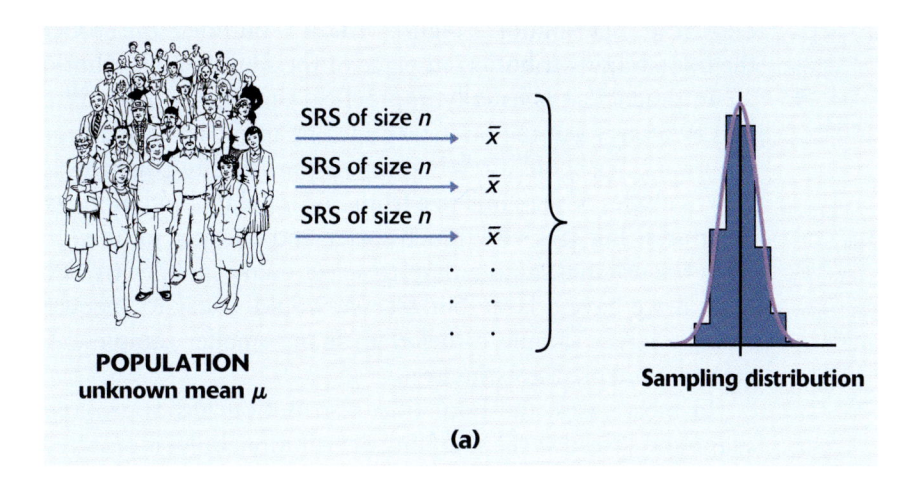

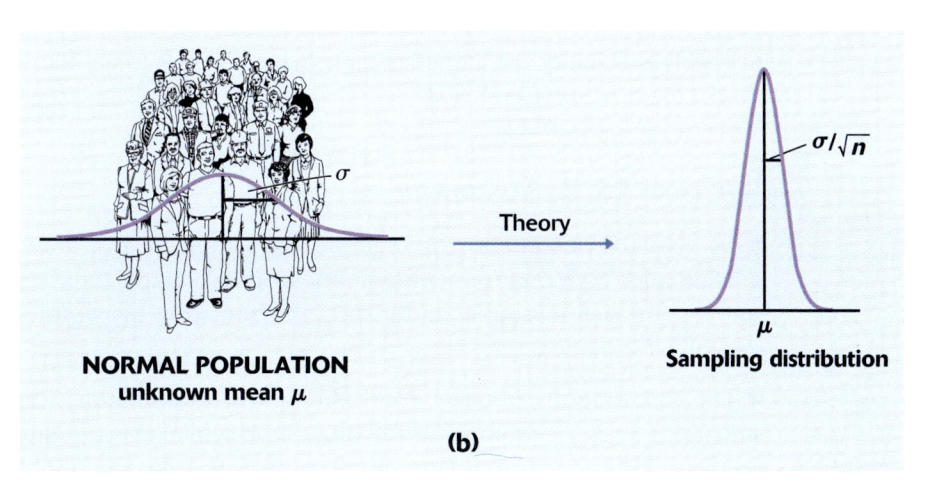

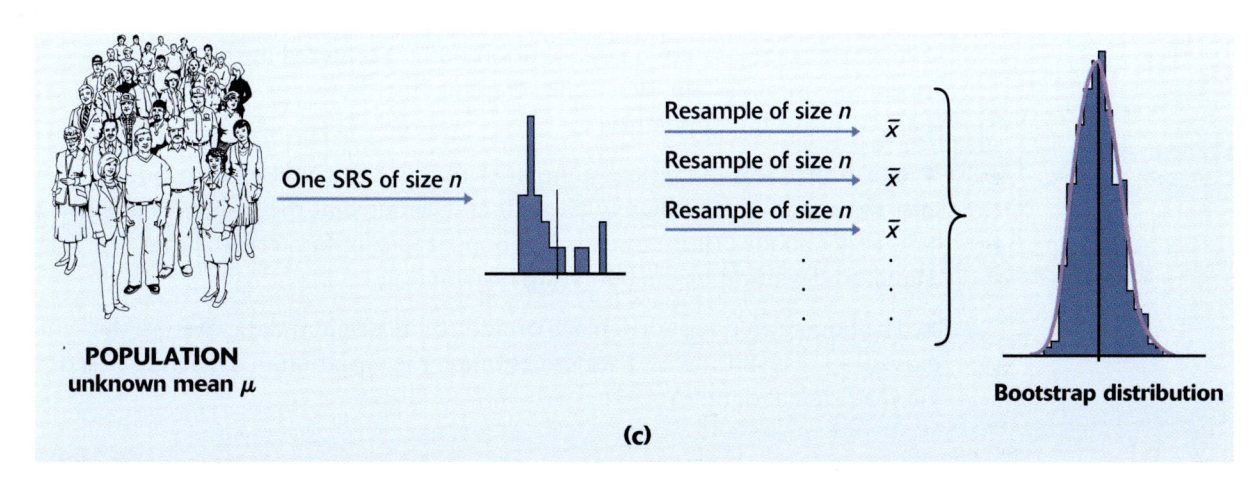

FIGURE 18.5 **(a)** The idea of the sampling distribution of the sample mean $\bar{x}$: take very many samples, collect the value of $\bar{x}$ from each, and look at the distribution of these values. **(b)** The probability theory shortcut: if we know that the population values follow a Normal distribution, theory tells us that the sampling distribution of $\bar{x}$ is also Normal. **(c)** The bootstrap idea: when theory fails and we can afford only one sample, that sample stands in for the population and the distribution of $\bar{x}$ in many resamples stands in for the sampling distribution.

bootstrap distribution. Figure 18.5(c) outlines the process. Then use the bootstrap distribution in place of the sampling distribution.

In practice, it is usually impractical to actually draw all possible resamples. We carry out the bootstrap idea by using 1000 or so randomly chosen resamples. We could directly estimate the sampling distribution by choosing 1000 samples of the same size from the original population, as Figure 18.5(a) illustrates. But it is very much faster and cheaper to let software resample from the original sample than to select many samples from the population. Even if we have a large budget, we would prefer to spend it on obtaining a single larger sample rather than many smaller samples. A larger sample gives a more precise estimate.

In most cases, the bootstrap distribution has approximately the same shape and spread as the sampling distribution, but it is centered at the original statistic value rather than the parameter value. The bootstrap allows us to calculate standard errors for statistics for which we don't have formulas and to check Normality for statistics that theory doesn't easily handle. We'll do this in the next section.

Section 18.2 Summary

- To bootstrap a statistic (for example, the sample mean), draw hundreds of **resamples** with replacement from the original sample data, calculate the statistic for each resample, and inspect the **bootstrap distribution** of the resampled statistics.

- The bootstrap distribution approximates the sampling distribution of the statistic. This is an example of the **plug-in principle**: use a quantity based on the sample to approximate a similar quantity from the population.

- Bootstrap distributions usually have approximately the same shape and spread as the sampling distribution but are centered at the statistic (from the original data) when the sampling distribution is centered at the parameter (of the population).

- Use graphs and numerical summaries to determine whether the bootstrap distribution is approximately Normal and centered at the original statistic and to get an idea of its spread. The **bootstrap standard error** is the standard deviation of the bootstrap distribution.

- The bootstrap does not replace or add to the original data. We use the bootstrap distribution as a way to estimate the variation in a statistic based on the original data.

Section 18.2 Exercises

Unless an exercise instructs you otherwise, use 1000 resamples for all bootstrap exercises. S-PLUS uses 1000 resamples unless you ask for a different number. Always save your bootstrap results in a file or S-PLUS object, as in Example 8.3, so that you can use them again later.

18.3 Spending by shoppers. Here are the dollar amounts spent by 50 consecutive shoppers at a supermarket. We are willing to regard this as an SRS of all shoppers at this market.

3.11	8.88	9.26	10.81	12.69	13.78	15.23	15.62	17.00	17.39
18.36	18.43	19.27	19.50	19.54	20.16	20.59	22.22	23.04	24.47
24.58	25.13	26.24	26.26	27.65	28.06	28.08	28.38	32.03	34.98
36.37	38.64	39.16	41.02	42.97	44.08	44.67	45.40	46.69	48.65
50.39	52.75	54.80	59.07	61.22	70.32	82.70	85.76	86.37	93.34

(a) Make a histogram of the data. The distribution is slightly skewed.

(b) The central limit theorem says that the sampling distribution of the sample mean $\bar{x}$ becomes Normal as the sample size increases. Is the sampling distribution roughly Normal for $n = 50$? To find out, bootstrap these data and inspect the bootstrap distribution of the mean.

18.4 Guinea pig survival times. The lifetimes of machines before a breakdown and the survival times of cancer patients after treatment are typically strongly right-skewed. Here are the survival times (in days) of 72 guinea pigs in a medical trial:[4]

43	45	53	56	56	57	58	66	67	73
74	79	80	80	81	81	81	82	83	83
84	88	89	91	91	92	92	97	99	99
100	100	101	102	102	102	103	104	107	108
109	113	114	118	121	123	126	128	137	138
139	144	145	147	156	162	174	178	179	184
191	198	211	214	243	249	329	380	403	511
522	598								

(a) Make a histogram of the survival times. The distribution is strongly skewed.

(b) The central limit theorem says that the sampling distribution of the sample mean $\bar{x}$ becomes Normal as the sample size increases. Is the sampling distribution roughly Normal for $n = 72$? To find out, bootstrap these data and inspect the bootstrap distribution of the mean (use a Normal quantile plot). How does the distribution differ from Normality? Is the bootstrap distribution more or less skewed than the data distribution?

18.5 More on supermarket shoppers. Here is an SRS of 10 of the amounts spent from Exercise 18.3:

18.43	52.75	50.39	34.98	19.27	19.54	15.23	17.39	12.69	93.34

We expect the sampling distribution of $\bar{x}$ to be less close to Normal for samples of size 10 than for samples of size 50 from a skewed distribution. This sample includes a high outlier.

(a) Create and inspect the bootstrap distribution of the sample mean from these data. Is it less close to Normal than your distribution from Exercise 18.3?

(b) Compare the bootstrap standard errors for your two runs. What accounts for the larger standard error for the smaller sample?

18.6 **More on survival times.** Here is an SRS of 20 of the guinea pig survival times from Exercise 18.4:

| 92 | 123 | 88 | 598 | 100 | 114 | 89 | 522 | 58 | 191 |
| 137 | 100 | 403 | 144 | 184 | 102 | 83 | 126 | 53 | 79 |

We expect the sampling distribution of $\bar{x}$ to be less close to Normal for samples of size 20 than for samples of size 72 from a skewed distribution. These data include some extreme high outliers.

(a) Create and inspect the bootstrap distribution of the sample mean for these data. Is it less close to Normal than your distribution from Exercise 18.4?

(b) Compare the bootstrap standard errors for your two runs. What accounts for the larger standard error for the smaller sample?

18.7 **Comparing standard errors.** We have two ways to estimate the standard deviation of a sample mean $\bar{x}$: use the formula $s/\sqrt{n}$ for the standard error or use the bootstrap standard error. Find the sample standard deviation s for the 50 amounts spent in Exercise 18.3 and use it to find the standard error $s/\sqrt{n}$ of the sample mean. How closely does your result agree with the bootstrap standard error from your resampling in Exercise 18.3?

18.3 Bootstrap Distributions and Standard Errors

In this section we'll use the bootstrap procedure to find bootstrap distributions and standard errors for statistics other than the mean. The shape of the bootstrap distribution approximates the shape of the sampling distribution, so we can use the bootstrap distribution to check Normality of the sampling distribution. If the sampling distribution appears to be Normal and centered at the true parameter value, we can use the bootstrap standard error to calculate a t confidence interval. So we need to use the bootstrap to check the center of the sampling distribution as well as the shape and spread. It turns out that the bootstrap does not reveal the center directly, but rather reveals the *bias*.

> **BIAS**
>
> A statistic used to estimate a parameter is **biased** when its sampling distribution is not centered at the true value of the parameter. The bias of a statistic is the mean of the sampling distribution minus the parameter.
>
> The bootstrap method allows us to check for bias by seeing whether the bootstrap distribution of a statistic is centered at the statistic of the original random sample. The bootstrap estimate of bias is the mean of the bootstrap distribution minus the statistic for the original data.

18.3 Bootstrap Distributions and Standard Errors

TABLE 18.1	Selling prices (in $1000) for an SRS of 50 Seattle real estate sales in 2002								
142	232	132.5	200	362	244.95	335	324.5	222	225
175	50	215	260	307	210.95	1370	215.5	179.8	217
197.5	146.5	116.7	449.9	266	265	256	684.5	257	570
149.4	155	244.9	66.407	166	296	148.5	270	252.95	507
705	1850	290	164.95	375	335	987.5	330	149.95	190

CASE 18.2 — REAL ESTATE SALE PRICES

We are interested in the sales prices of residential property in Seattle. Unfortunately, the data available from the county assessor's office do not distinguish residential property from commercial property. Most of the sales in the assessor's records are residential, but a few large commercial sales in a sample can greatly increase the mean selling price. We prefer to use a measure of center that is more resistant than the mean. When we do this, we know less about the sampling distribution than if we used the mean to measure center. The bootstrap is very handy in such settings.

Table 18.1 gives the selling prices for a random sample (SRS of size 50) from the population of all 2002 Seattle real estate sales, as recorded by the county assessor. The sales include houses, condominiums, and commercial real estate but exclude plots of undeveloped land.[5]

Figure 18.6 describes these data with a histogram and Normal quantile plot. As we expect, the distribution is strongly skewed to the right. There are several high outliers, which may be commercial sales.

EXAMPLE 18.4 — Bootstrapping the mean selling price

CASE 18.2

The skewness of the distribution of real estate prices affects the sampling distribution of the sample mean. We cannot see the sampling distribution directly without taking many samples, but the bootstrap distribution gives us a clue. Figure 18.7 shows the bootstrap distribution of the sample mean $\bar{x}$ based on 1000 resamples from the data in Table 18.1. The distribution is skewed to the right—that is, a sample of size 50 is not large enough to allow us to act as if $\bar{x}$ has a Normal distribution.

There is some good news as well. The bootstrap distribution shows that the outliers do not cause large bias—the mean of the bootstrap distribution is approximately equal to the sample mean of the data in Table 18.1 (the solid and dotted lines nearly coincide). We conclude that the sampling distribution is skewed but has small bias. This isn't surprising: we know that $\bar{x}$ is an unbiased estimator of the population mean μ, whether or not the population has a Normal distribution.

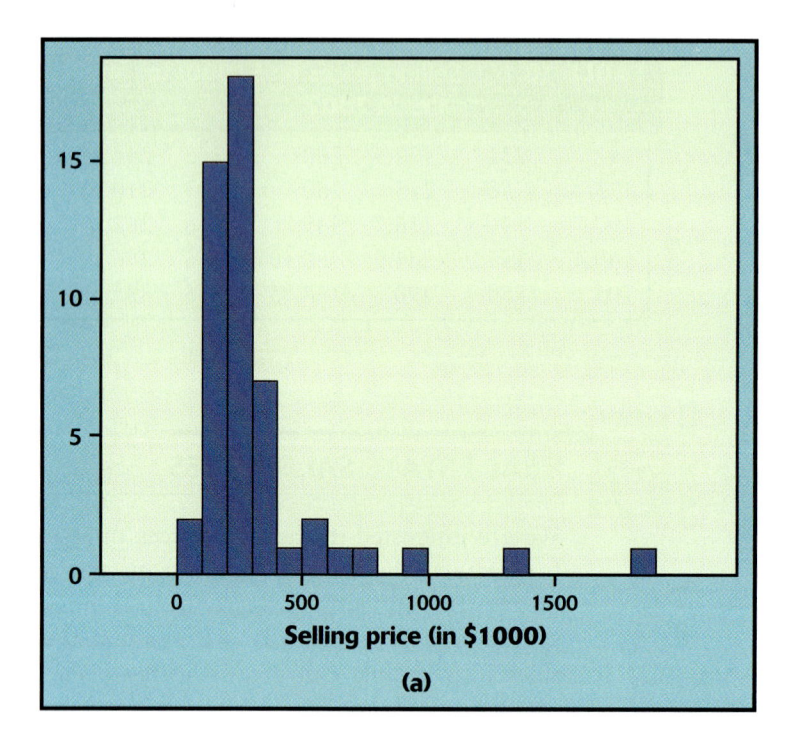

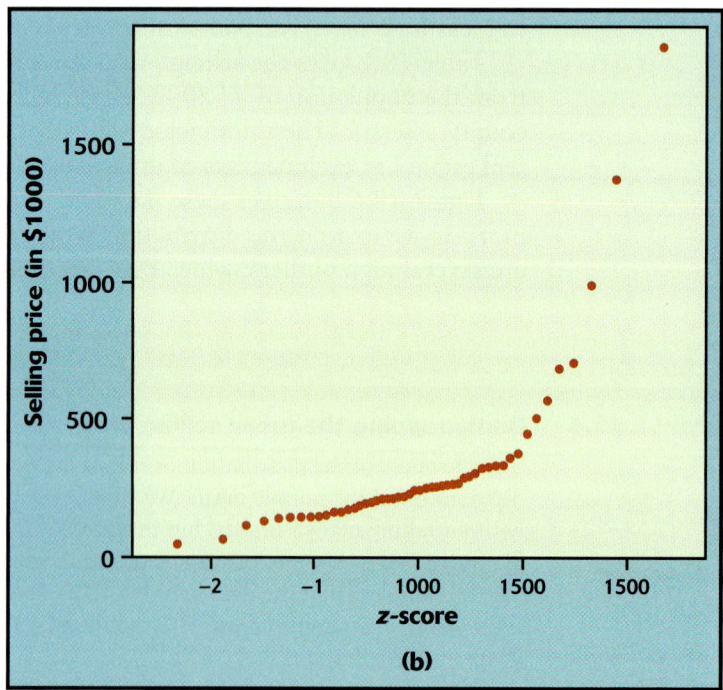

FIGURE 18.6 Graphical displays of the 50 selling prices in Table 18.1. The distribution is strongly skewed, with high outliers.

18.3 Bootstrap Distributions and Standard Errors

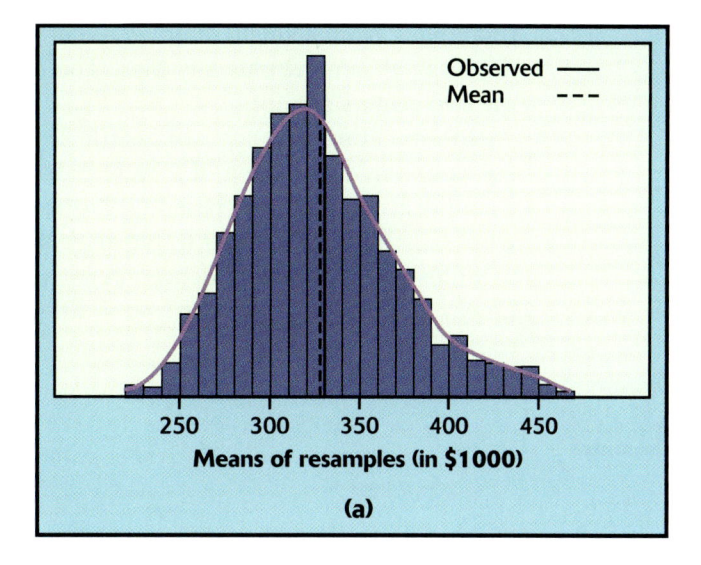

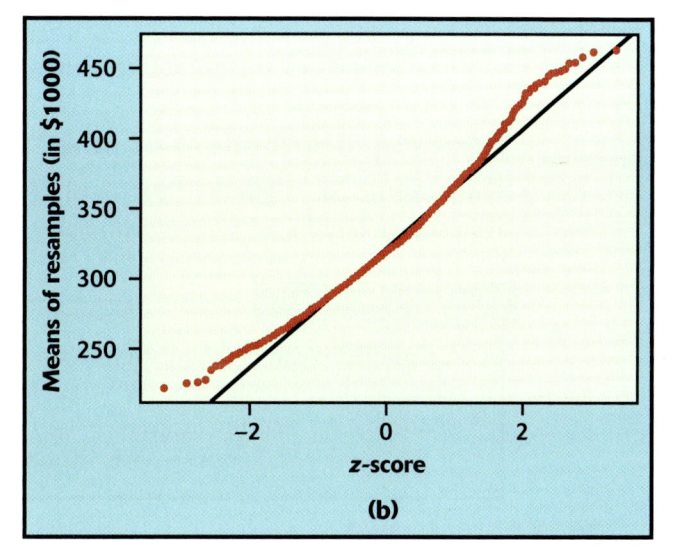

FIGURE 18.7 The bootstrap distribution of the sample means of 1000 resamples from the data in Table 18.1. The bootstrap distribution is right-skewed, so we conclude that the sampling distribution of $\bar{x}$ is right-skewed as well.

The conclusion of Example 18.4 is based on the following principle.

> **BOOTSTRAP DISTRIBUTIONS AND SAMPLING DISTRIBUTIONS**
>
> For most statistics, bootstrap distributions approximate the shape, spread, and bias of the actual sampling distribution.

Bootstrap distributions differ from the actual sampling distributions in the location of their centers. The sampling distribution of a statistic used to estimate a parameter is centered at the actual value of the parameter in the population, plus any bias. The bootstrap distribution, generated by resampling from a single sample, is centered at the value of the statistic for the original sample, plus any bias. The two biases are similar even though the two centers are not.

In estimating the center of Seattle real estate prices, we cannot act as if the sampling distribution of $\bar{x}$ were Normal. We have two alternatives: use a confidence interval not based on Normality or choose a measure of center whose distribution is closer to Normal. We will see that advanced bootstrap methods do produce confidence intervals not based on Normality. For now, however, we choose to bootstrap a different statistic that is more resistant to skewness and outliers.

APPLY YOUR KNOWLEDGE

18.8 **Supermarket shoppers.** What is the bootstrap estimate of the bias from your resamples in Exercise 18.3? What does this tell you about the bias encountered in using $\bar{x}$ to estimate the mean spending for all shoppers at this market?

18.9 **Guinea pig survival.** What is the bootstrap estimate of the bias from your resamples in Exercise 18.4? What does this tell you about the bias encountered in using $\bar{x}$ to estimate the mean survival time for all guinea pigs that receive the same experimental treatment?

Bootstrap distributions of other statistics

One statistic we might consider in place of the mean is the median. Here, instead, we'll use a *25% trimmed mean*.

> **TRIMMED MEAN**
>
> A **trimmed mean** is the mean of only the center observations in a data set. In particular, the 25% trimmed mean $\bar{x}_{25\%}$ ignores the smallest 25% and the largest 25% of the observations. It is the mean of the middle 50% of the observations.

Recall that the median is the mean of the 1 or 2 middle observations. The trimmed mean often does a better job of representing the average of typical observations than does the median. Bootstrapping trimmed means also works better than bootstrapping medians, because the bootstrap doesn't work well for statistics that depend on only 1 or 2 observations.

EXAMPLE 18.5 **25% trimmed mean for the real estate data**

CASE 18.2

We don't need any distribution facts about the trimmed mean to use the bootstrap. We bootstrap the 25% trimmed mean just as we bootstrapped the sample mean: draw 1000 resamples, calculate the 25% trimmed mean for each resample, and form the bootstrap distribution from these 1000 values. Figure 18.8 shows the result.

Comparing Figures 18.7 and 18.8 shows that the bootstrap distribution of the trimmed mean is less skewed than the bootstrap distribution of the mean and is closer to Normal. It is close enough that we will calculate a confidence interval for the population trimmed mean based on Normality. (If high accuracy were

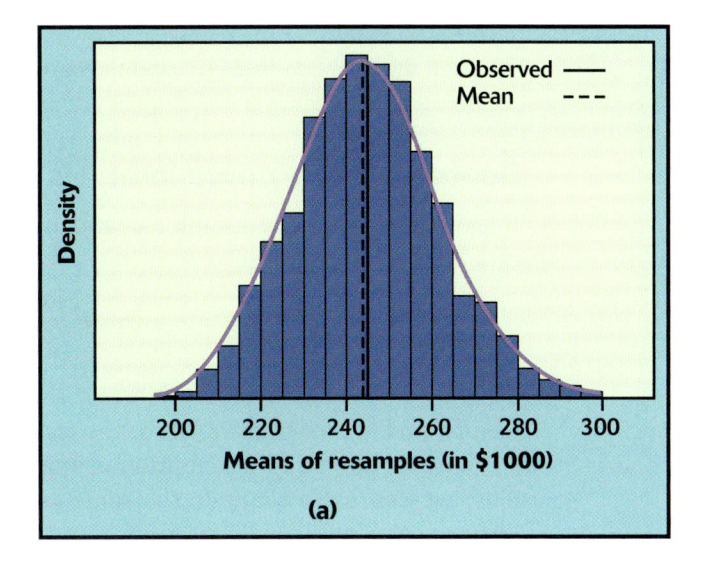

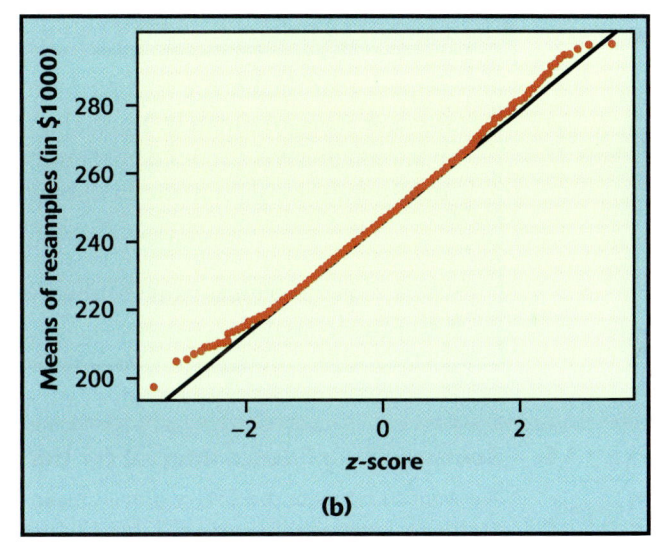

18.3 Bootstrap Distributions and Standard Errors

FIGURE 18.8 The bootstrap distribution of the 25% trimmed means of 1000 resamples from the data in Table 18.1. The bootstrap distribution is roughly Normal.

important, we would prefer one of the more accurate confidence interval procedures we discuss later.)

The distribution of the trimmed mean is also narrower than that of the mean. For a long-tailed distribution such as this, the 25% trimmed mean is a less variable estimate of the center of the population than is the ordinary mean. Here is the summary output from S-PLUS:

```
Number of Replications: 1000

Summary Statistics:
         Observed   Mean    Bias      SE
TrimMean      244  244.7  0.7171   16.83
```

The trimmed mean for the sample is $\bar{x}_{25\%} = 244$, the mean of the 1000 trimmed means of the resamples is 244.7, and the bootstrap standard error is 16.83.

Bootstrap *t* confidence intervals

Recall the familiar one-sample *t* confidence interval (page 435) for the mean of a Normal population,

$$\bar{x} \pm t^* \frac{s}{\sqrt{n}}$$

This interval is based on the Normal sampling distribution of the sample mean $\bar{x}$ and the formula $s/\sqrt{n}$ for the standard error of $\bar{x}$.

When a bootstrap distribution is approximately Normal and has small bias, we can use essentially the same recipe with the bootstrap standard error to get a confidence interval for any parameter.

> **BOOTSTRAP *t* CONFIDENCE INTERVAL***
>
> Suppose that the bootstrap distribution of a statistic from an SRS of size *n* is approximately Normal and that the bias is small. An approximate level *C* confidence interval for the parameter that corresponds to this statistic by the plug-in principle is
>
> $$\text{statistic} \pm t^* \text{SE}_{\text{boot, statistic}}$$
>
> where t^* is the critical value of the $t(n-1)$ distribution with area *C* between $-t^*$ and t^*.

EXAMPLE 18.6 Bootstrap *t* confidence interval for the trimmed mean

CASE 18.2

We want to estimate the 25% trimmed mean of the population of all 2002 Seattle real estate selling prices. Table 18.1 gives an SRS of size $n = 50$. The software output in Example 18.5 shows that the trimmed mean of this sample is $\bar{x}_{25\%} = 244$ and that the bootstrap standard error of this statistic is $\text{SE}_{\text{boot}, \bar{x}_{25\%}} = 16.83$. A 95% confidence interval for the population trimmed mean is therefore

$$\bar{x}_{25\%} \pm t^* \text{SE}_{\text{boot}, \bar{x}_{25\%}} = 244 \pm (2.009)(16.83)$$
$$= 244 \pm 33.81$$
$$= (210.19, 277.81)$$

Because Table D does not have entries for $n - 1 = 49$ degrees of freedom, we used $t^* = 2.009$, the entry for 50 degrees of freedom.

We are 95% confident that the 25% trimmed mean (the mean of the middle 50%) for the population of real estate sales in Seattle in 2002 is between $210,190 and $277,810.

---

*There is another "bootstrap *t* confidence interval" in common use. It estimates the value of t^* that is appropriate for the data rather than using a value from a *t* table.

APPLY YOUR KNOWLEDGE

18.10 Confidence interval for shoppers' mean spending. Your investigation in Exercise 18.3 found that the bootstrap distribution of the mean is reasonably Normal and has small bias.

(a) What is the bootstrap t 95% confidence interval for the population mean μ, based on your resamples from Exercise 18.3?

(b) Also find the standard one-sample t confidence interval. The two intervals differ only in the standard errors used. How similar are the intervals?

18.11 Trimmed mean for shoppers' spending. Because the distribution of amounts spent by supermarket shoppers (Exercise 18.3) is strongly skewed, we might choose to use a measure of center more resistant than the mean.

(a) Find the 25% trimmed mean for this sample of 50 shoppers. Why is the trimmed mean smaller than the mean?

(b) Use the bootstrap t method to give a 95% confidence interval for the 25% trimmed mean spending in the population of all shoppers.

18.12 Median for shoppers' spending. We remarked that bootstrap methods often work poorly for the median. Construct and inspect the bootstrap distribution of the median for resamples from the shopper spending data (Exercise 18.3). Present a plot of the distribution and explain carefully why you would not use the bootstrap t confidence interval for the population median.

Bootstrapping to compare two groups

Two-sample problems (Section 7.2) are among the most common statistical settings. In a two-sample problem, we wish to compare two populations, such as male and female customers, based on separate samples from each population. When both populations are roughly Normal, the two-sample t procedures compare the two population means. The bootstrap can also compare two populations, without the Normality condition and without the restriction to comparison of means. The most important new idea is that bootstrap resampling must mimic the "separate samples" design that produced the original data.

BOOTSTRAP FOR COMPARING TWO POPULATIONS

Given independent SRSs of sizes n and m from two populations:

1. Draw a resample of size n with replacement from the first sample and a separate resample of size m from the second sample. Compute a statistic that compares the two groups, such as the difference between the two sample means.

2. Repeat this resampling process hundreds of times.

3. Construct the bootstrap distribution of the statistic. Inspect its shape, bias, and bootstrap standard error in the usual way.

EXAMPLE 18.7

Service times in telecommunications

Incumbent local exchange carriers (ILECs), such as Verizon, install and maintain local telephone lines, lease capacity, and perform repairs for the competing local exchange carriers (CLECs). Figure 18.9 shows density curves and Normal quantile plots for the repair times (in hours) of 1664 service requests from customers of Verizon and 23 requests from customers of a CLEC during the same time period. The distributions are both far from Normal. Here are some summary statistics:

Service provider	n	$\bar{x}$	s
Verizon	1664	8.4	14.7
CLEC	23	16.5	19.5
Difference		−8.1	

The data suggest that repair times may be longer for the CLEC. The mean repair time, for example, is almost twice as long for CLEC customers as for Verizon customers.

In the setting of Example 18.7 we want to estimate the difference of population means, $\mu_1 - \mu_2$, but we are reluctant to use the two-sample t confidence interval because one of the samples is both quite small and very skewed. To compute the bootstrap standard error for the difference in sample means $\bar{x}_1 - \bar{x}_2$, resample separately from the two samples. Each of our 1000 resamples consists of two group resamples, one of size 1664 drawn with replacement from the Verizon data and one of size 23 drawn with replacement from the CLEC data. For each combined resample, compute the statistic $\bar{x}_1 - \bar{x}_2$. The 1000 differences form the bootstrap distribution. The bootstrap standard error is the standard deviation of the bootstrap distribution. Here is the S-PLUS output:

```
Number of Replications: 1000

Summary Statistics:
          Observed    Mean     Bias      SE
meanDiff   -8.098   -8.251  -0.1534   4.052
```

The bootstrap distribution and Normal quantile plot are shown in Figure 18.10. The bootstrap distribution is not close to Normal. It has a short right tail and a long left tail, so that it is skewed to the left. We are unwilling to use a bootstrap t confidence interval. That is, no method based on Normality is safe. In Section 18.5, we will see that there are other ways of using the bootstrap to get confidence intervals that can be safely used in this and similar examples.

18.3 Bootstrap Distributions and Standard Errors

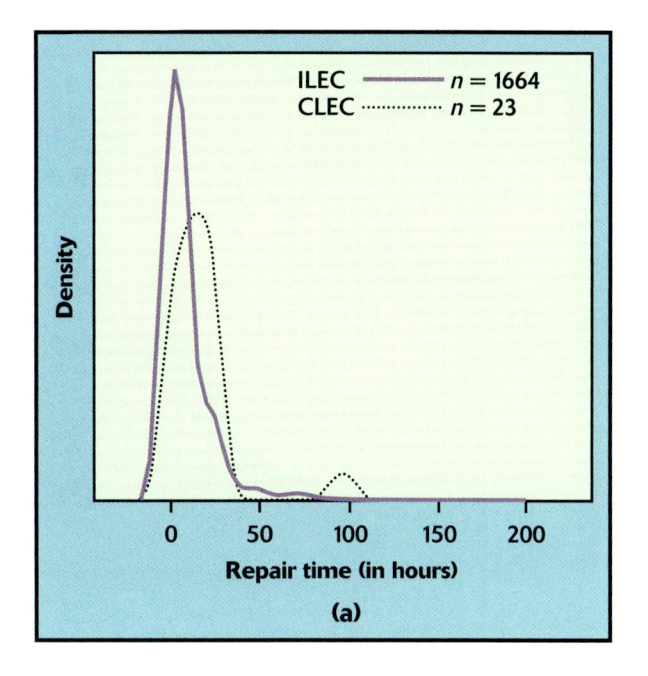

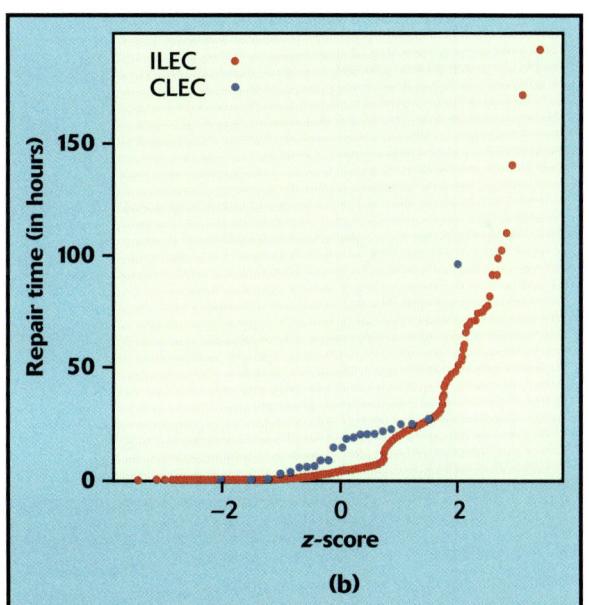

FIGURE 18.9 Comparing the distributions of repair times (in hours) for 1664 requests from Verizon customers and 23 requests for customers of a CLEC. The top panel shows density curves and the bottom panel shows Normal quantile plots. (The density curves appear to show negative repair times—this is due to how the density curves are calculated from data, not because any times are negative.)

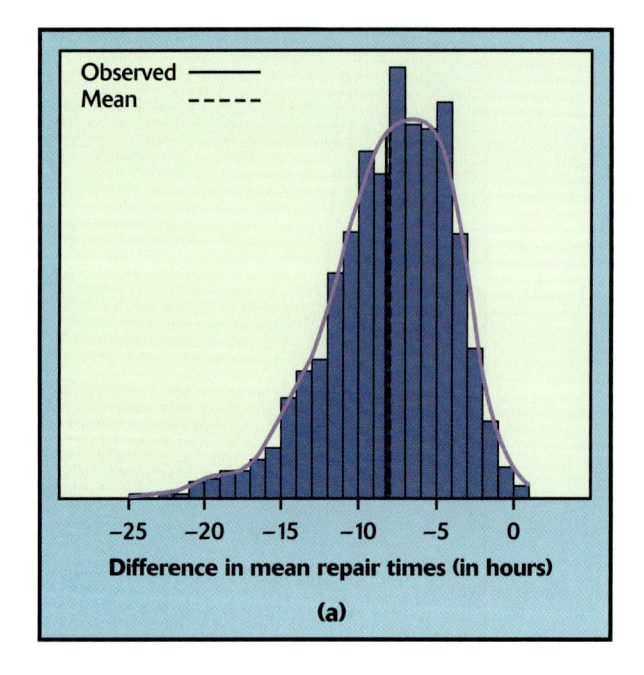

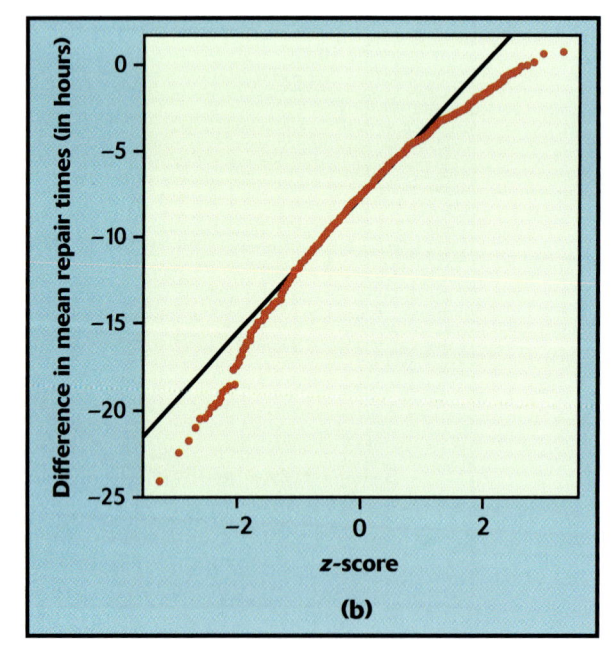

FIGURE 18.10 The bootstrap distribution of the difference in means for the Verizon and CLEC repair time data.

APPLY YOUR KNOWLEDGE

18.13 **Compare standard errors.** The formula for the standard error of $\bar{x}_1 - \bar{x}_2$ is $\sqrt{s_1^2/n_1 + s_2^2/n_2}$ (see page 464). This formula does not depend on Normality. How does this formula-based standard error for the data of Example 18.7 compare with the bootstrap standard error?

18.14 **An experiment in education.** Table 7.3 (page 465) gives the scores on a test of reading ability for two groups of third-grade students. The treatment group used "directed reading activities" and the control group followed the same curriculum without the activities.

(a) Bootstrap the difference in means $\bar{x}_1 - \bar{x}_2$ and report the bootstrap standard error.

(b) Inspect the bootstrap distribution. Is a bootstrap t confidence interval appropriate? If so, give the interval.

(c) Compare the bootstrap results with the two-sample t confidence interval reported on page 478.

18.15 **Healthy versus failed companies.** Table 7.4 (page 476) contains the ratio of current assets to current liabilities for random samples of healthy firms and failed firms. Find the difference in means (healthy minus failed).

(a) Bootstrap the difference in means $\bar{x}_1 - \bar{x}_2$ and look at the bootstrap distribution. Does it meet the conditions for a bootstrap t confidence interval?

(b) Report the bootstrap standard error and the bootstrap t confidence interval.

(c) Compare the bootstrap results with the two-sample t confidence interval reported on page 479.

BEYOND THE BASICS: THE BOOTSTRAP FOR A SCATTERPLOT SMOOTHER

The bootstrap idea can be applied to quite complicated statistical methods, such as the scatterplot smooth illustrated in Chapter 2 (page 126).

EXAMPLE 18.8

Do some lottery numbers pay more?

The New Jersey Pick-It Lottery is a daily numbers game run by the state of New Jersey. We'll analyze the first 254 drawings after the lottery was started in 1975.[6] Buying a ticket entitles a player to pick a number between 000 and 999. Half of the money bet each day goes into the prize pool. (The state takes the other half.) The state picks a winning number at random, and the prize pool is shared equally among all winning tickets.

Although all numbers are equally likely to win, numbers chosen by fewer people have bigger payoffs if they win because the prize is shared among fewer tickets. Figure 18.11 is a scatterplot of the first 254 winning numbers and their payoffs. What patterns can we see?

The straight line in Figure 18.11 is the least-squares regression line. The line shows a general trend of higher payoffs for larger winning numbers. The curve in the figure was fitted to the plot by a scatterplot smoother that follows local patterns in the data rather than being constrained to a straight line. The curve suggests that there were larger payoffs for numbers in the intervals 000 to 100, 400 to 500, 600 to 700, and 800 to 999. When people pick "random" numbers, they tend to choose numbers starting with 2, 3, 5, or 7, so these numbers have lower payoffs. This pattern disappeared after 1976—it appears that players noticed the pattern and changed their number choices.

Are the patterns displayed by the scatterplot smooth just chance? We can use the bootstrap distribution of the smoother's curve to get an idea of how

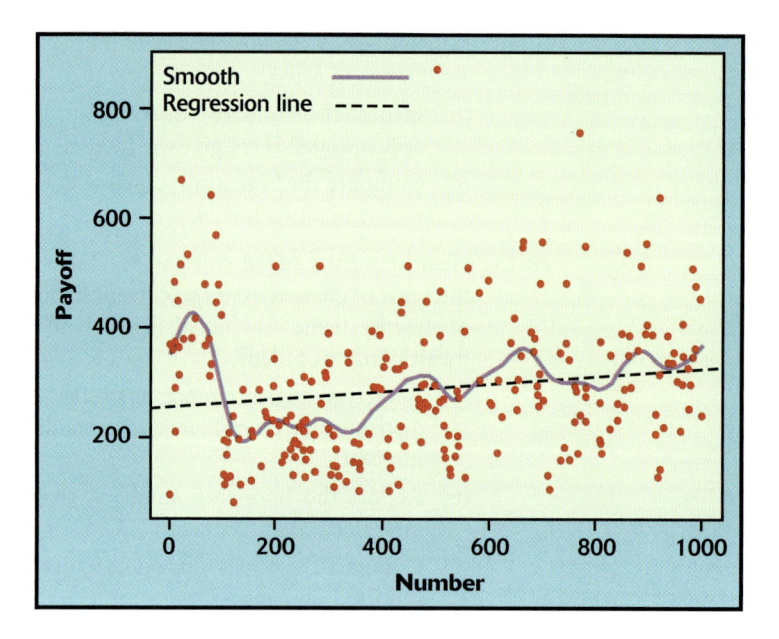

FIGURE 18.11 The first 254 winning numbers in the New Jersey Pick-It Lottery and the payoffs for each. To see patterns we use least-squares regression (*line*) and a scatterplot smoother (*curve*).

much random variability there is in the curve. Each resample "statistic" is now a curve rather than a single number. Figure 18.12 shows the curves that result from applying the smoother to 20 resamples from the 254 data points in Figure 18.11. The original curve is the thick line. The spread of the resample curves about the original curve shows the sampling variability of the output of the scatterplot smoother.

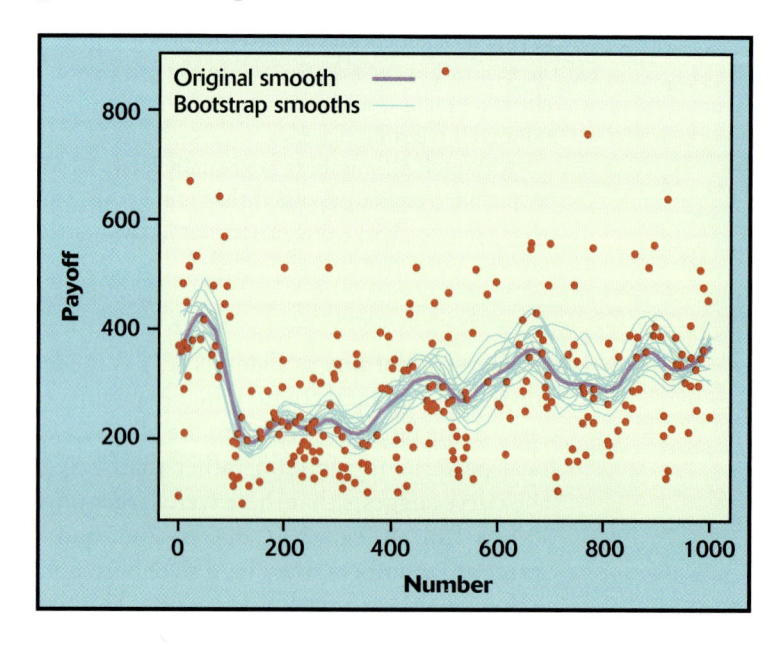

FIGURE 18.12 The curves produced by the scatterplot smoother for 20 resamples from the data displayed in Figure 18.11. The curve for the original sample is the heavy line.

Nearly all the bootstrap curves mimic the general pattern of the original smooth curve, showing, for example, the same low average payoffs for numbers in the 200s and 300s. This suggests that these patterns are real, not just chance.

SECTION 18.3 SUMMARY

- Bootstrap distributions mimic the shape, spread, and bias of sampling distributions.

- The **bootstrap standard error** is the standard deviation of the bootstrap distribution. It measures how much a statistic varies under random sampling.

- The bootstrap estimate of **bias** is the mean of the bootstrap distribution minus the statistic for the original data. Small bias means that the bootstrap distribution is centered at the statistic of the original sample and suggests that the sampling distribution of the statistic is centered at the population parameter.

- The bootstrap can estimate the sampling distribution, bias, and standard error of a wide variety of statistics, such as the **trimmed mean**.

- If the bootstrap distribution is approximately Normal and the bias is small, we can give a **bootstrap t confidence interval**, statistic $\pm t^*\text{SE}$, for the parameter. Do not use this t interval if the bootstrap distribution is not Normal or shows substantial bias.

- To bootstrap a statistic that compares two samples, such as the difference in sample means, we draw separate resamples from the two original samples.

SECTION 18.3 EXERCISES

18.16 **Standard error.** What is the difference between the standard deviation of a sample and the standard error of a statistic such as the sample mean?

18.17 **Seattle real estate sales: the mean.** Figure 18.7 shows one bootstrap distribution of the mean selling price for Seattle real estate in 2002. Repeat the resampling of the data in Table 18.1 to get another bootstrap distribution for the mean.

(a) Plot the bootstrap distribution and compare it with Figure 18.7. Although resamples are random, we expect 1000 resamples to always produce similar results. Are the two bootstrap distributions similar?

(b) Compare the bootstrap standard error of the mean to the bootstrap standard error of the 25% trimmed mean for the same data in Example 18.5. How do the two bootstrap distributions (Figures 18.7 and 18.8) reflect this comparison?

(c) Why should we *not* report a bootstrap t confidence interval for the mean?

18.18 Seattle real estate sales: the median. Bootstrap the median for the Seattle real estate sales data in Table 18.1.

(a) What is the bootstrap standard error of the median?

(b) Look at the bootstrap distribution of the median. Despite the small standard error, why might we not want to report a t confidence interval for the median?

18.19 Really Normal data. The following data are an SRS from the standard Normal distribution $N(0, 1)$, produced by a software Normal random number generator:

0.01	−0.04	−1.02	−0.13	−0.36	−0.03	−1.88	0.34	−0.00	1.21
−0.02	−1.01	0.58	0.92	−1.38	−0.47	−0.80	0.90	−1.16	0.11
0.23	2.40	0.08	−0.03	0.75	2.29	−1.11	−2.23	1.23	1.56
−0.52	0.42	−0.31	0.56	2.69	1.09	0.10	−0.92	−0.07	−1.76
0.30	−0.53	1.47	0.45	0.41	0.54	0.08	0.32	−1.35	−2.42
0.34	0.51	2.47	2.99	−1.56	1.27	1.55	0.80	−0.59	0.89
−2.36	1.27	−1.11	0.56	−1.12	0.25	0.29	0.99	0.10	0.30
0.05	1.44	−2.46	0.91	0.51	0.48	0.02	−0.54		

(a) Make a histogram and Normal quantile plot. Do the data appear to follow the $N(0, 1)$ distribution?

(b) Bootstrap the mean and report the bootstrap standard error.

(c) Why do your bootstrap results suggest that a t confidence interval is appropriate? Give the 95% bootstrap t interval.

18.20 CEO salaries. The following data are the salaries, including bonuses (in millions of dollars), for the chief executive officers (CEOs) of small companies in 1993.[7] Small companies are defined as those with annual sales greater than $5 million and less than $350 million.

145	621	262	208	362	424	339	736	291	58	498	643	390	332	750
368	659	234	396	300	343	536	543	217	298	198	406	254	862	204
206	250	21	298	350	800	726	370	536	291	808	543	149	350	242
1103	213	296	317	482	155	802	200	282	573	388	250	396	572	

(a) Display the data using a histogram and Normal quantile plot. Describe the shape, center, and spread of the distribution.

(b) Create the bootstrap distribution for the 25% trimmed mean or, if your software won't calculate trimmed means, the median.

(c) Is a bootstrap t confidence interval appropriate? If so, calculate the 95% interval.

18.21 Clothing for runners. Your company sells exercise clothing and equipment on the Internet. To design clothing, you collect data on the physical characteristics of your customers. Here are the weights in kilograms for a sample of 25 male runners. Assume these runners are a random sample of your potential male customers.

67.8	61.9	63.0	53.1	62.3	59.7	55.4	58.9	60.9
69.2	63.7	68.3	92.3	64.7	65.6	56.0	57.8	66.0
62.9	53.6	65.0	55.8	60.4	69.3	61.7		

Since your products are aimed toward the "average male," you are interested in seeing how much the subjects in your sample vary from the average weight.

(a) Calculate the sample standard deviation s for these weights.

(b) We have no formula for the standard error of s. Find the bootstrap standard error for s.

(c) What does the standard error indicate about how accurate the sample standard deviation is as an estimate of the population standard deviation?

(d) Would it be appropriate to give a bootstrap t interval for the population standard deviation? Why or why not?

18.22 Clothing for runners, interquartile range. If your software will calculate the interquartile range, repeat the previous exercise using the interquartile range in place of the standard deviation to measure spread.

18.23 Mortgage refusal rates. The Association of Community Organizations for Reform Now (ACORN) did a study on refusal rates in mortgage lending by 20 banks in major cities.[8] They recorded the percent of mortgage applications refused for both white and minority applicants. Here are the results for the 20 banks:

Bank	Minority	White	Bank	Minority	White
Harris Trust	20.9	3.7	Provident National	49.7	20.1
NCNB Texas	23.2	5.5	Worthen	44.6	19.1
Crestar	23.1	6.7	Hibernia National	36.4	16.0
Mercantile	30.4	9.0	Sovron	32.0	16.0
First NB Commerce	42.7	13.9	Bell Federal	10.6	5.6
Texas Commerce	62.2	20.6	Security Pacific Arizona	34.3	18.4
Comerica	39.5	13.4	Core States	42.3	23.3
First of America	38.4	13.2	Citibank Arizona	26.5	15.6
Boatman's National	26.2	9.3	Manufacturers Hanover	51.5	32.4
First Commercial	55.9	21.0	Chemical	47.2	29.7

ACORN is concerned that minority applicants are refused more often than are white applicants.

(a) Display the data by making separate histograms and Normal quantile plots for the minority and white refusal rates. Is there anything in the displays to indicate that the sampling distribution of the difference in means might be non-Normal?

(b) Give a two-sample paired t 95% confidence interval for the difference in the population means. What do your results show?

(c) Bootstrap the difference in means $\bar{x}_1 - \bar{x}_2$. (You should resample banks rather than resampling the minority and white refusal rates separately. Or you could compute the difference in refusal rates for each bank, and resample the differences.) Does the bootstrap distribution indicate that a t confidence interval is appropriate? If yes, give a 95% t confidence interval using the bootstrap standard error. How does your result compare with the traditional interval in (b)?

18.24 Billionaires. Each year, the business magazine *Forbes* publishes a list of the world's billionaires. In 2002, the magazine found 497 billionaires. Here is the wealth, as estimated by *Forbes* and rounded to the nearest $100 million, of an SRS of 20 of these billionaires:[9]

8.6	1.3	5.2	1.0	2.5	1.8	2.7	2.4	1.4	3.0
5.0	1.7	1.1	5.0	2.0	1.4	2.1	1.2	1.5	1.0

You are interested in (vaguely) "the wealth of typical billionaires." Bootstrap an appropriate statistic, inspect the bootstrap distribution, and draw conclusions based on this sample.

18.25 **Seeking the source of the skew.** Why is the bootstrap distribution of the difference in mean Verizon and CLEC repair times in Figure 18.10 so skewed? Let's investigate by bootstrapping the mean of the CLEC data and comparing it with the bootstrap distribution for the mean for Verizon customers. CASE 18.1

(a) Bootstrap the mean for the CLEC data. Compare the bootstrap distribution with the bootstrap distribution of the Verizon repair times in Figure 18.3.

(b) Given what you see in part (a), what is the source of the skew in the bootstrap distribution of the difference in means $\bar{x}_1 - \bar{x}_2$?

18.4 How Accurate Is a Bootstrap Distribution?*

The sampling distribution of a statistic displays the variation in the statistic due to selecting samples at random from the population. We understand that the statistic will vary from sample to sample, so that inference about the population must take this random variation into account. For example, the margin of error in a confidence interval expresses the uncertainty due to sampling variation. Now we have used the bootstrap distribution as a substitute for the sampling distribution. We thus introduce another source of random variation: resamples are chosen at random from the original sample.

> **SOURCES OF VARIATION IN A BOOTSTRAP DISTRIBUTION**
>
> Bootstrap distributions and conclusions based on them include two sources of random variation:
>
> 1. The original sample is chosen at random from the population.
>
> 2. Bootstrap resamples are chosen at random from the original sample.

Figure 18.13 shows the entire process. The population distribution (top left) has two peaks and is clearly not close to Normal. Below the figure are histograms of five random samples from this population, each of size 50. The sample means $\bar{x}$ are marked on each histogram. These vary from sample to sample. The distribution of the $\bar{x}$-values from all possible samples

---
*This section is optional.

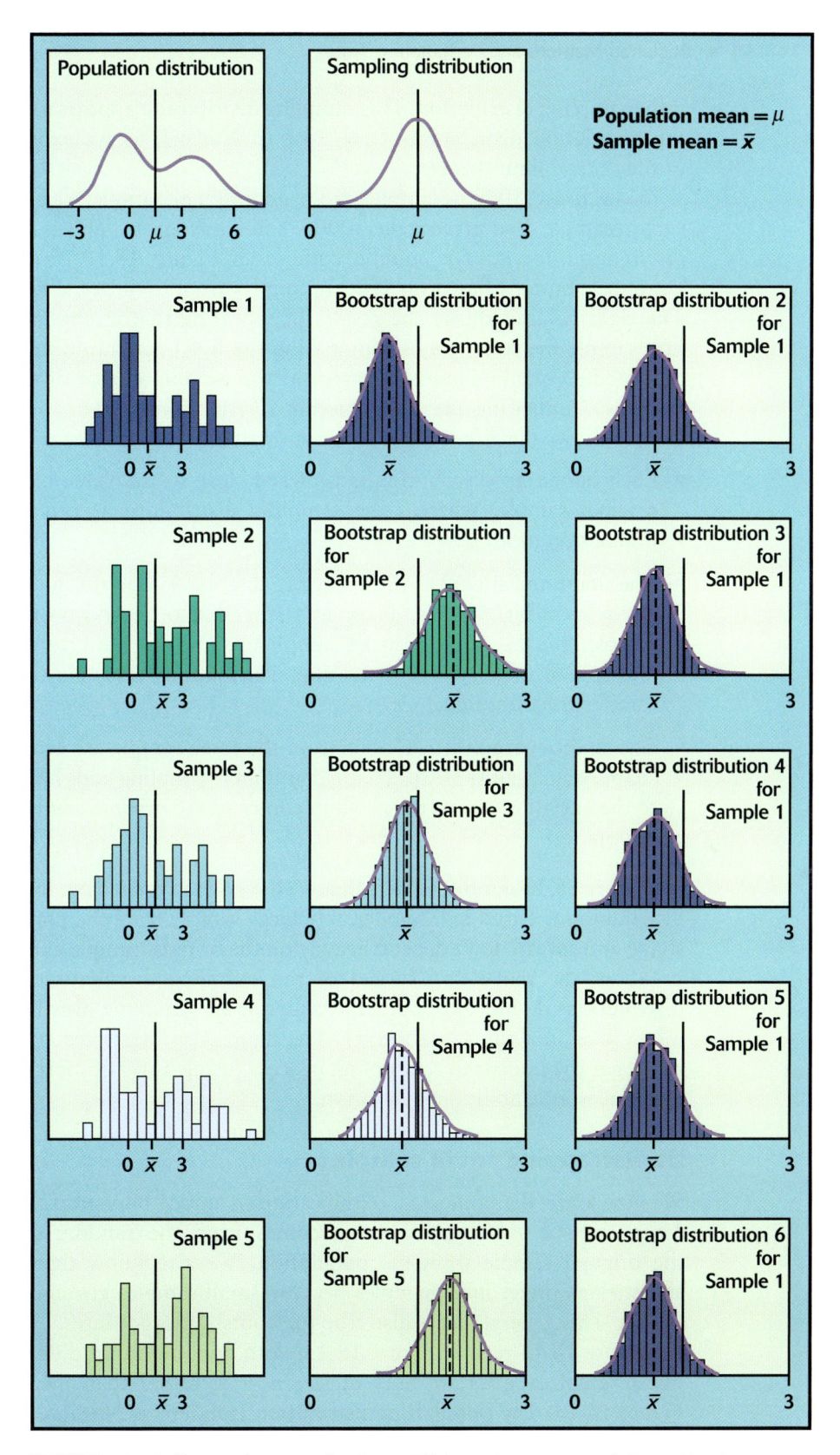

FIGURE 18.13 Five random samples ($n = 50$) from the same population, with a bootstrap distribution for the sample mean formed by resampling from each of the five samples. At the right are five more bootstrap distributions from the first sample. In all cases, the mean of the bootstrap distribution is nearly indistinguishable from $\bar{x}$, so is not shown separately.

is the sampling distribution. This sampling distribution appears to the right of the population distribution. It is close to Normal, as we expect because of the central limit theorem.

Now draw 1000 resamples from an original sample, calculate $\bar{x}$ for each resample, and present the 1000 $\bar{x}$'s in a histogram. This is a bootstrap distribution for $\bar{x}$. The middle column in Figure 18.13 displays bootstrap distributions based on 1000 resamples from each of the five samples. The right column shows the results of repeating the resampling from the first sample five more times. Comparing the five bootstrap distributions in the middle column shows the effect of the random choice of the original samples. Comparing the six bootstrap distributions drawn from the first sample shows the effect of the random resampling. Here's what we see:

- Each bootstrap distribution is centered close to the value of $\bar{x}$ from its original sample, whereas the sampling distribution is centered at the population mean μ.

- The shape and spread of the bootstrap distributions in the middle column also vary a bit. That is, shape and spread also depend on the original sample, but the variation from sample to sample is not great. The shape and spread of all of the bootstrap distributions resemble those of the sampling distribution.

- The six bootstrap distributions from the same sample are very similar in shape, center, and spread. That is, random resampling adds little variation to the variation due to the random choice of the original sample from the population.

Figure 18.13 reinforces facts that we have already relied on. If a bootstrap distribution is based on a moderately large sample from the population, its shape and spread don't depend heavily on the original sample and do inform us about the shape and spread of the sampling distribution. Bootstrap distributions do not have the same center as the sampling distribution; they mimic bias, not the actual center. The figure also illustrates an important new fact: the bootstrap resampling process (using 1000 or more resamples) introduces little additional variation.

Bootstrapping small samples

We now know that almost all of the variation among bootstrap distributions for a statistic such as the mean comes from the random selection of the original sample from the population. We also know that in general statisticians prefer large samples because small samples give more variable results. This general fact is also true for bootstrap procedures.

Figure 18.14 repeats Figure 18.13, with two important differences. The five original samples are only of size $n = 9$, rather than the $n = 50$ of Figure 18.13. The population distribution (top left) is Normal, so that the sampling distribution of $\bar{x}$ is Normal despite the small sample size. The bootstrap distributions in the middle column show more variation in shape and spread than those for larger samples in Figure 18.13. Notice, for example, how the skewness of the fourth sample produces a skewed bootstrap

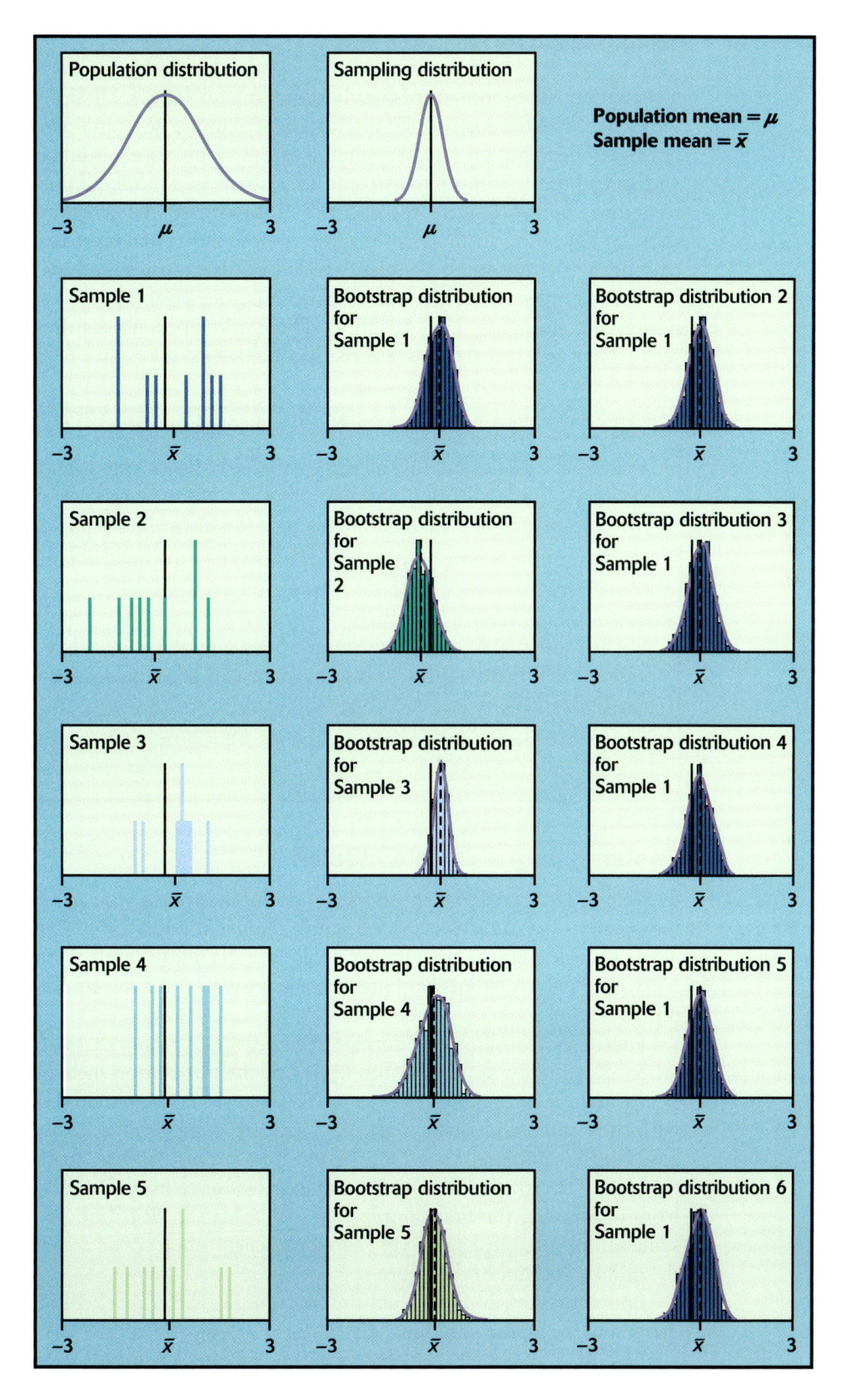

FIGURE 18.14 Five random samples ($n = 9$) from the same population, with a bootstrap distribution for the sample mean formed by resampling from each of the five samples. At the right are five more bootstrap distributions from the first sample. In all cases, the mean of the bootstrap distribution is nearly indistinguishable from $\bar{x}$, so is not shown separately.

distribution. The bootstrap distributions are no longer all similar to the sampling distribution at the top of the column. We can't trust that a bootstrap distribution from so small a sample will closely mimic the shape and spread of the sampling distribution. Bootstrap confidence intervals will sometimes be too long or too short, or too long in one direction and too short in the other. In most cases these errors tend to balance out, but they may not with very small samples. The six bootstrap distributions based on the first sample are again very similar. Because we used 1000 resamples, resampling still adds little variation. There are subtle effects that can't be seen from a few pictures, but the main conclusions are clear.

> **DEALING WITH VARIATION IN BOOTSTRAP DISTRIBUTIONS**
>
> For most statistics, almost all the variation in bootstrap distributions comes from the selection of the original sample from the population. You can reduce this variation by using a larger original sample.
>
> Bootstrapping does not overcome the weakness of small samples as a basis for inference. Some bootstrap procedures (we will discuss BCa and tilting later) are usually more accurate than standard methods, but even they may not be accurate for very small samples. Use caution in any inference—including bootstrap inference—from a small sample.
>
> The bootstrap resampling process using 1000 or more resamples introduces little additional variation.

Bootstrapping a sample median

In Section 18.3 we chose to bootstrap the 25% trimmed mean rather than the median. We did this in part because the usual bootstrapping procedure doesn't work well for the median unless the original sample is quite large. Now we will try bootstrapping the median in order to understand the difficulties.

Figure 18.15 follows the format of Figures 18.13 and 18.14. The population distribution appears at top left, with the population median marked. Below in the left column are five samples of size $n = 15$ from this population, with their sample medians marked. Bootstrap distributions for the median based on resampling from each of the five samples appear in the middle column. The right column again displays five more bootstrap distributions from resampling the first sample. The six bootstrap distributions from the same sample are once again very similar to each other, so we concentrate on the middle column in the figure.

Bootstrap distributions from the five samples differ markedly from each other and from the sampling distribution at the top of the column. The median of a resample can only be one of the 15 observations in the original sample and is usually one of the few in the middle. Each bootstrap distribution repeats the same few values. The sampling distribution, on the other hand, contains the medians of all possible samples and is not

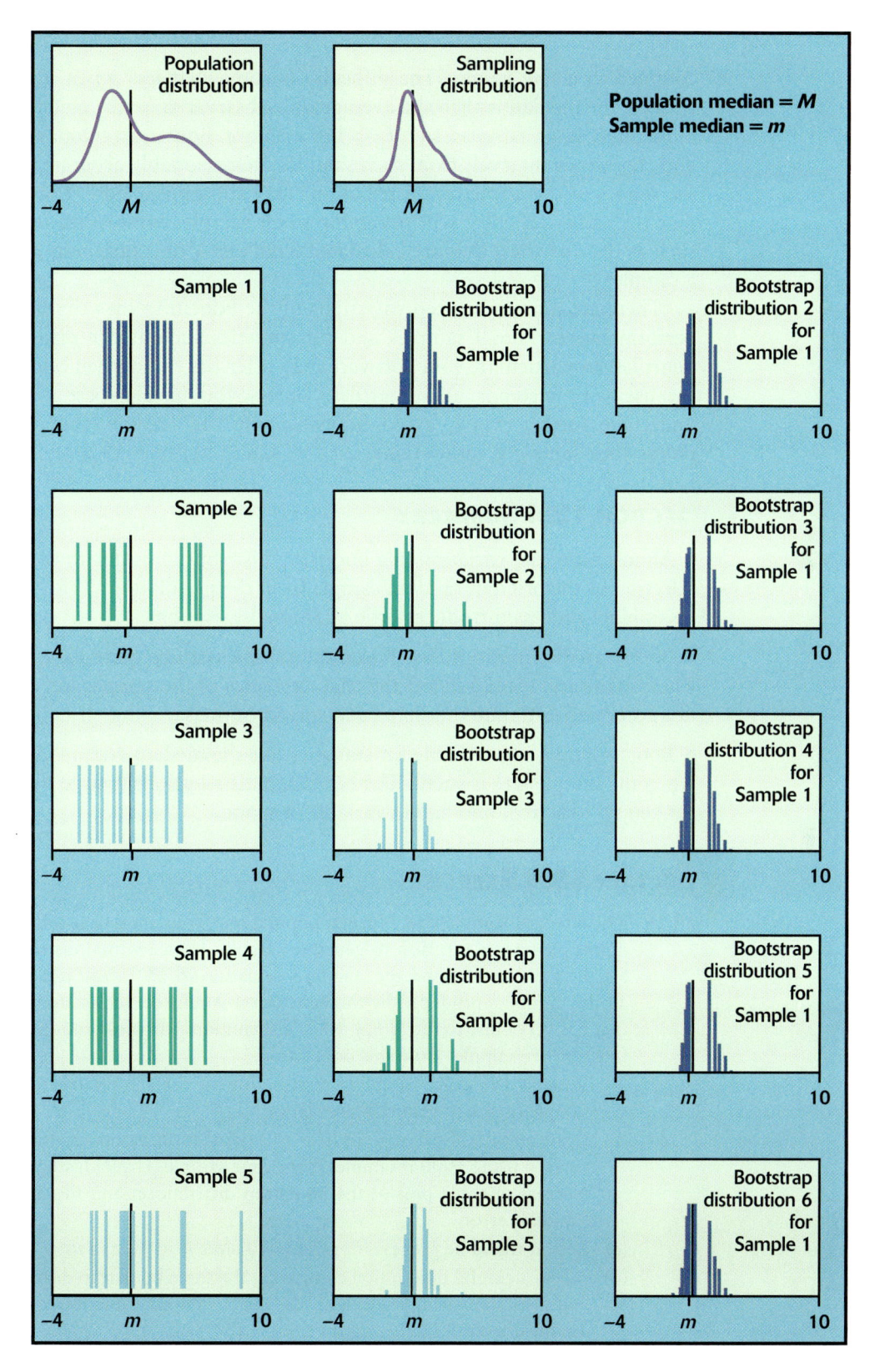

FIGURE 18.15 Five random samples ($n = 15$) from the same population, with a bootstrap distribution for the sample median formed by resampling from each of the five samples. At the right are five more bootstrap distributions from the first sample.

confined to a few values. The difficulty is somewhat less when *n* is even, because the median is then the average of 2 observations. It is much less for moderately large samples, say, $n = 100$ or more. Bootstrap standard errors and confidence intervals from such samples are reasonably accurate, though the shapes of the bootstrap distributions may still appear odd. You can see that the same difficulty will occur for small samples with other statistics, such as the quartiles, that are calculated from just 1 or 2 observations from a sample.

There are more advanced variations of the bootstrap idea that improve performance for small samples and for statistics such as the median and quartiles. In particular, your software may offer the "smoothed bootstrap" for use with medians and quartiles. Unless you have expert advice or undertake further study, avoid bootstrapping the median and quartiles unless your sample is rather large.

SECTION 18.4 SUMMARY

■ Almost all of the variation in a bootstrap distribution is due to the selection of the original random sample from the population. The resampling process introduces little additional variation.

■ Bootstrap distributions based on small samples can be quite variable. Their shape and spread reflect the characteristics of the sample and may not accurately estimate the shape and spread of the sampling distribution.

■ Bootstrapping is unreliable for statistics like the median and quartiles when the sample size is small. The bootstrap distributions tend to be broken up (discrete) and highly variable in shape.

SECTION 18.4 EXERCISES

18.26 **The effect of sample size: Normal population.** Your statistical software no doubt includes a function to generate samples from Normal distributions. Set the mean to 8.4 and the standard deviation to 14.7. You can think of all the numbers produced by this function if it ran forever as a population that has very close to the $N(8.4, 14.7)$ distribution. Samples produced by the function are samples from this population.
  (a) What is the exact sampling distribution of the sample mean $\bar{x}$ for a sample of size *n* from this population?
  (b) Draw an SRS of size $n = 10$ from this population. Bootstrap the sample mean $\bar{x}$ using 1000 resamples from your sample. Give a histogram and Normal quantile plot of the bootstrap distribution and the bootstrap standard error.
  (c) Repeat the same process for samples of sizes $n = 40$ and $n = 160$.
  (d) Write a careful description comparing the three bootstrap distributions and also comparing them with the exact sampling distribution. What are the effects of increasing the sample size?

18.27 **The effect of sample size: non-Normal population.** The data for Example 18.7 include 1664 repair times for customers of Verizon, the local telephone company in their area. In that example these observations formed a sample. Now we will treat these

1664 observations as a population. The population distribution appears in Figure 18.9. The population mean is $\mu = 8.4$, and the population standard deviation is $\sigma = 14.7$.

(a) Although we don't know the shape of the sampling distribution of the sample mean $\bar{x}$ for a sample of size n from this population, we do know the mean and standard deviation of this distribution. What are they?

(b) Draw an SRS of size $n = 10$ from this population. Bootstrap the sample mean $\bar{x}$ using 1000 resamples from your sample. Give a histogram and Normal quantile plot of the bootstrap distribution and the bootstrap standard error.

(c) Repeat the same process for samples of sizes $n = 40$ and $n = 160$.

(d) Write a careful description comparing the three bootstrap distributions. What are the effects of increasing the sample size?

18.28 **Normal versus non-Normal populations.** The populations in the two previous exercises have the same mean and standard deviation, but one is very close to Normal and the other is strongly non-Normal. Based on your work in these exercises, how does non-Normality of the population affect the bootstrap distribution of $\bar{x}$? How does it affect the bootstrap standard error? Do either of these effects diminish when we start with a larger sample? Explain what you have observed based on what you know about the sampling distribution of $\bar{x}$ and the way in which bootstrap distributions mimic the sampling distribution.

18.5 Bootstrap Confidence Intervals

To this point, we have met just one type of inference procedure based on resampling: bootstrap t confidence intervals. We can calculate a bootstrap t confidence interval for any parameter by bootstrapping the corresponding statistic (the plug-in principle). We don't need conditions on the population or special knowledge about the sampling distribution of the statistic. The flexible and almost automatic nature of bootstrap t intervals is wonderful—but there is a catch. These intervals work well only when the bootstrap distribution tells us that the sampling distribution is approximately Normal and has small bias. How can we know whether these conditions are met well enough to trust the confidence interval? And what can we do if we don't trust the bootstrap t interval? This section deals with these important questions. We'll learn a quick way to check t confidence intervals for accuracy and learn alternative ways to calculate confidence intervals when t intervals aren't accurate.

Bootstrap percentiles as a check

Confidence intervals are based on the sampling distribution of a statistic. A 95% confidence interval starts by marking off the central 95% of the sampling distribution. The t critical values in any t confidence interval are a shortcut to marking off this central 95%. The shortcut requires special conditions that are not always met, so t intervals are not always appropriate. One way to check whether t intervals (using either bootstrap or formula standard errors) are reasonable is therefore to compare them with the central 95% of the bootstrap distribution. The 2.5th and 97.5th percentiles mark off the central 95%. The interval between the 2.5th and 97.5th percentiles of the bootstrap distribution is often used as a confidence interval in its own right. It is known as a *bootstrap percentile confidence interval*.

> **BOOTSTRAP PERCENTILE CONFIDENCE INTERVALS**
>
> The interval between the 2.5th and 97.5th percentiles of the bootstrap distribution of a statistic is a 95% **bootstrap percentile confidence interval** for the corresponding parameter.
>
> If the bias of the bootstrap distribution is small and the distribution is close to Normal, the bootstrap t and percentile confidence intervals will agree closely. If they do not agree, this is evidence that the Normality and bias conditions are not met. Neither type of interval should be used if this is the case.

EXAMPLE 18.9 — **Seattle real estate sales: the trimmed mean**

In Examples 18.5 and 18.6 we found a 95% bootstrap t confidence interval for the 25% trimmed mean, but we also noted that the bootstrap distribution was a bit skewed. We'd like to know how that affects the accuracy of the t confidence interval.

The S-PLUS bootstrap output includes the 2.5th and 97.5th percentiles of the bootstrap distribution. Using these, the percentile interval for the trimmed mean of the Seattle real estate sales is 213.1 to 279.4. This is quite close to the bootstrap t interval 210.2 to 277.8 found in Example 18.6. This suggests that both intervals are reasonably accurate.

The bootstrap t interval for the trimmed mean of real estate sales is

$$\overline{x}_{25\%} \pm t^* \text{SE}_{\text{boot}, \overline{x}_{25\%}} = 244 \pm 33.81$$

We can learn something by also writing the percentile interval starting at the statistic $\overline{x}_{25\%} = 244$. In this form, it is

$$244.0 - 30.9, \quad 244.0 + 35.4$$

Unlike the t interval, the percentile interval is not symmetric—its endpoints are different distances from the statistic. The slightly greater distance to the 97.5th percentile reflects the slight right-skewness of the bootstrap distribution.

APPLY YOUR KNOWLEDGE

18.29 Percentile confidence intervals. What percentiles of the bootstrap distribution are the endpoints of a 90% bootstrap percentile confidence interval?

18.30 IQ scores of seventh-grade students. The following data are the IQ scores for 78 seventh-grade students at a middle school.[10] We will treat these data as a random sample of all seventh-grade IQ scores in the region.

111	102	128	123	93	105	107	91	118	124	72	110
100	114	113	126	111	107	107	114	120	116	103	103
114	103	132	127	123	77	115	106	111	119	79	98
111	105	124	97	119	90	97	113	127	86	110	96
100	109	128	102	110	112	112	108	136	110	107	112
104	113	106	120	74	114	89	130	118	103	105	93
104	128	119	115	112	106						

We expect the distribution of IQ scores to be approximately Normal. The sample size is reasonably large, so the sampling distribution of the mean should be close to Normal.

(a) Make a Normal quantile plot of the data. Is the distribution approximately Normal?

(b) Use the formula $s/\sqrt{n}$ to find the standard error of the mean. Give the 95% t confidence interval based on this standard error.

(c) Bootstrap the mean of the IQ scores. Make a histogram and Normal quantile plot of the bootstrap distribution. Does the bootstrap distribution appear Normal? What is the bootstrap standard error? Give the bootstrap t 95% confidence interval.

(d) Give the 95% percentile confidence interval. How well do your three confidence intervals agree? Was bootstrapping needed to find a reasonable confidence interval, or was the formula confidence interval good enough?

Confidence intervals for the correlation coefficient

The bootstrap allows us to find standard errors and confidence intervals for a wide variety of statistics. We have to this point done this for the mean, the trimmed mean, the difference of means, and (with less success) the median. Now we will bootstrap the correlation coefficient. This is our first use of the bootstrap for a statistic that depends on two related variables.

CASE 18.3

BASEBALL SALARIES AND PERFORMANCE

Major League Baseball (MLB) owners claim they need direct or indirect limitations on player salaries to maintain competitiveness among richer and poorer teams. This argument assumes that higher salaries are needed to attract better players. Is there a relationship between an MLB player's salary and his performance?

Table 18.2 contains the names, 2002 salaries, and career batting averages of 50 randomly selected MLB players (excluding pitchers).[11] The scatterplot in Figure 18.16 suggests that the relationship between salary and batting average is weak to nonexistent. The correlation is positive but small, $r = 0.107$. We wonder if this is significantly greater than 0. To find out, we can calculate a 95% confidence interval and see whether or not it covers 0.

TABLE 18.2　Major League Baseball salaries and batting averages

Name	Salary	Average	Name	Salary	Average
Matt Williams	$9,500,000	.269	Greg Colbrunn	$1,800,000	.307
Jim Thome	8,000,000	.282	Dave Martinez	1,500,000	.276
Jim Edmonds	7,333,333	.327	Einar Diaz	1,087,500	.216
Fred McGriff	7,250,000	.259	Brian L. Hunter	1,000,000	.289
Jermaine Dye	7,166,667	.240	David Ortiz	950,000	.237
Edgar Martinez	7,086,668	.270	Luis Alicea	800,000	.202
Jeff Cirillo	6,375,000	.253	Ron Coomer	750,000	.344
Rey Ordonez	6,250,000	.238	Enrique Wilson	720,000	.185
Edgardo Alfonzo	6,200,000	.300	Dave Hansen	675,000	.234
Moises Alou	6,000,000	.247	Alfonso Soriano	630,000	.324
Travis Fryman	5,825,000	.213	Keith Lockhart	600,000	.200
Kevin Young	5,625,000	.238	Mike Mordecai	500,000	.214
M. Grudzielanek	5,000,000	.245	Julio Lugo	325,000	.262
Tony Batista	4,900,000	.276	Mark L. Johnson	320,000	.207
Fernando Tatis	4,500,000	.268	Jason LaRue	305,000	.233
Doug Glanville	4,000,000	.221	Doug Mientkiewicz	285,000	.259
Miguel Tejada	3,625,000	.301	Jay Gibbons	232,500	.250
Bill Mueller	3,450,000	.242	Corey Patterson	227,500	.278
Mark McLemore	3,150,000	.273	Felipe Lopez	221,000	.237
Vinny Castilla	3,000,000	.250	Nick Johnson	220,650	.235
Brook Fordyce	2,500,000	.208	Thomas Wilson	220,000	.243
Torii Hunter	2,400,000	.306	Dave Roberts	217,500	.297
Michael Tucker	2,250,000	.235	Pablo Ozuna	202,000	.333
Eric Chavez	2,125,000	.277	Alexis Sanchez	202,000	.301
Aaron Boone	2,100,000	.227	Abraham Nunez	200,000	.224

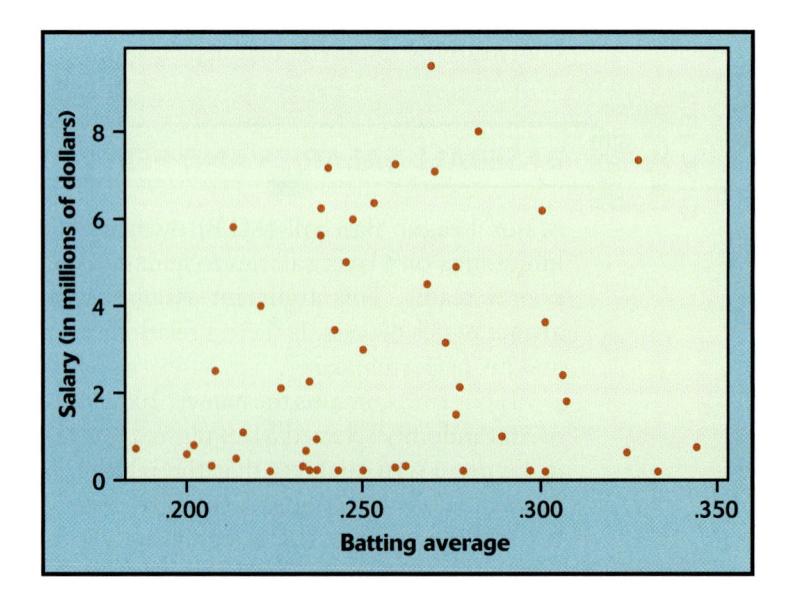

FIGURE 18.16 Batting average and salary for a random sample of 50 Major League Baseball players.

EXAMPLE 18.10

CASE 18.3

Bootstrapping the correlation

We use the same bootstrap procedures to find a confidence interval for the correlation coefficient as for other statistics. There is one point to note: because each observation consists of the batting average and salary for one player, we resample players (that is, observations). Resampling batting averages and salaries separately would lose the tie between a player's batting average and his salary.

Figure 18.17 shows the bootstrap distribution and Normal quantile plot for the sample correlation for 1000 resamples from the 50 players in our sample. The bootstrap distribution is reasonably Normal and has small bias, so a 95% bootstrap t confidence interval appears reasonable.

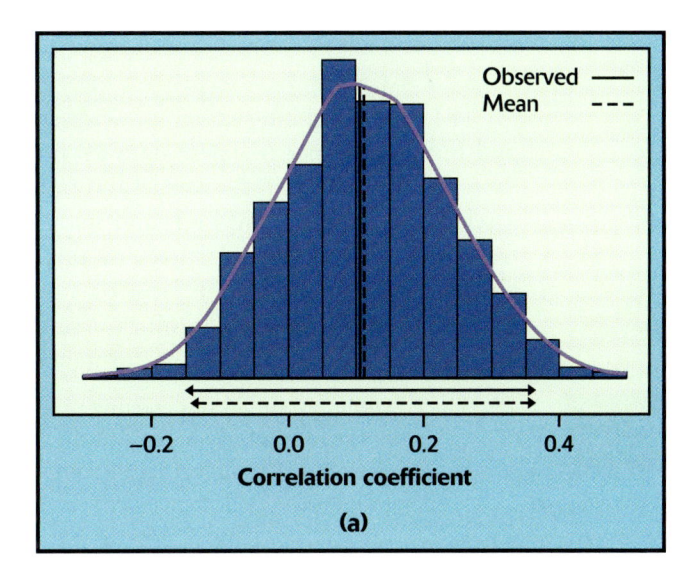

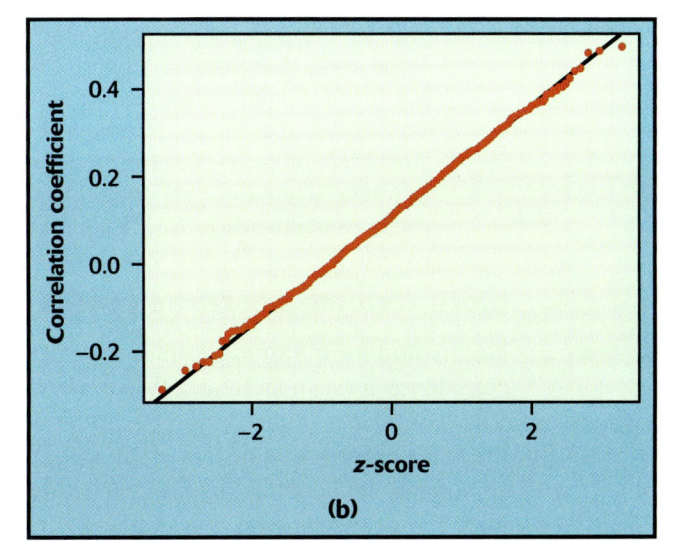

FIGURE 18.17 The bootstrap distribution and Normal quantile plot for the correlation r for 1000 resamples from the baseball player data in Table 18.2. The solid double-ended arrow below the distribution is the t interval, and the dashed arrow is the percentile interval.

The bootstrap standard error is $SE_{boot, r} = 0.125$. The t interval using the bootstrap standard error is

$$r \pm t^* SE_{boot, r} = 0.107 \pm (2.01)(0.125)$$
$$= 0.107 \pm 0.251$$
$$= (-0.144, 0.358)$$

The bootstrap percentile interval is

$$(2.5\text{th percentile}, 97.5\text{th percentile}) = (0.107 - 0.235, 0.107 + 0.249)$$
$$= (-0.128, 0.356)$$

The two confidence intervals are in reasonable agreement.

The confidence intervals give a wide range for the population correlation, and both include zero. These data do not provide significant evidence that there is a relationship between salary and batting average. There may be a relationship that could be found with a larger data set, but the evidence from this data set suggests that any relationship is fairly weak. Of course, batting average is only one facet of a player's performance. It is possible that we would discover a significant salary-performance relationship if we included several measures of performance.

APPLY YOUR KNOWLEDGE

18.31 Percentiles as an aid in detecting non-Normality. It is difficult to see any significant asymmetry in the bootstrap distribution of the correlation of Example 18.10. Compare the percentiles and the t interval; does the difference between these suggest any skewness? **CASE 18.3**

18.32 Wages and length of service. Table 10.1 (page 587) reports the wages and length of service for a random sample of 59 women who hold customer service jobs in Indiana banks. In Example 10.4, using a test that assumes a jointly Normal distribution for these variables, we found a highly significant relationship between wages and length of service. We may prefer inference that is not based on a Normal model. Bootstrap the correlation for these data. Give the bootstrap t and bootstrap percentile confidence intervals for the population correlation. Are these intervals trustworthy here? What do you conclude about the population? **CASE 10.1**

More accurate bootstrap confidence intervals

No method for obtaining confidence intervals produces exactly the intended confidence level in practice. When we compute what is supposed to be a 95% confidence interval, our method may give intervals that in fact capture the true parameter value less often, say, 92% or 85% of the time. Or instead of missing 2.5% of the time on each side, the method may in some settings miss 1% of the time on one side and 4% of the time on the other, giving a biased picture of where the parameter is.

accurate   We say that a method for obtaining 95% confidence intervals is **accurate** in a particular setting if 95% of the time it produces intervals that capture the parameter and if the 5% misses are shared equally between high and low misses. Confidence intervals are never exactly accurate because the conditions under which they work are never exactly satisfied in practice. The traditional t intervals, although reasonably robust, are affected by lack of Normality in the sampling distribution of the sample mean, especially skewness. Although the central limit theorem tells us that the sampling distribution of the mean becomes nearly Normal as the size of the sample increases, the effect of a skewed population can persist in the sampling distribution even for quite large samples.

One advantage of the bootstrap is that it allows us to check for skewness in a sampling distribution by inspecting the bootstrap distribution. We can also compare the bootstrap t and bootstrap percentile confidence intervals. When the sampling distribution is skewed, the percentile interval is shifted in the direction of the skewness, relative to the t interval. The intervals in both Example 18.9 and Example 18.10 reveal some right-skewness, though not enough to invalidate inference. The t and percentile intervals may not be sufficiently accurate when

- the statistic is strongly biased, as indicated by the bias estimate from the bootstrap,

- the sampling distribution of the statistic is clearly skewed, as indicated by the bootstrap distribution and by comparing the t and percentile intervals, or

- high accuracy is needed because the stakes are high (large sums of money or public welfare).

Bootstrap tilting and BCa intervals

Most confidence interval procedures are more accurate for larger sample sizes. The problem with t and percentile procedures is that they improve only slowly—they require 100 times more data to improve accuracy by a factor of 10—and so tend not to be very accurate except for quite large sample sizes. There are several bootstrap procedures that improve faster, requiring only 10 times more data to improve accuracy by a factor of 10. These procedures are quite accurate unless the sample size is very small. The **bootstrap bias-corrected accelerated (BCa)** and **bootstrap tilting** methods are accurate in a wide variety of settings, have reasonable computation requirements (by modern standards), and do not produce excessively wide intervals.

BCa
bootstrap tilting

These procedures are not as intuitively clear as the t and percentile methods, which is why we did not meet them earlier. Now that you understand the bootstrap, however, you should always use one of these more accurate methods if your software offers them.

EXAMPLE 18.11

CASE 18.2

Seattle real estate sales: the mean

The 2002 Seattle real estate sales data are strongly skewed (Figure 18.6), and the skewness persists in the sampling distribution of the mean (Figure 18.7). Generally, we prefer resistant measures of center such as the trimmed mean or median for skewed data. However, the mean is easily understood by the public and is needed for some purposes, such as projecting taxes based on total sales value.

```
Number of Replications: 1000

Summary Statistics:
        Observed  Mean   Bias    SE
mean    329.3    326.9  -2.383  43.9

Percentiles:
       2.5%   5.0%   95.0%  97.5%
mean   252.5  261.7  408.3  433.1

BCa Confidence Limits:
       2.5%  5%     95%    97.5%
mean   270   279.6  442.7  455.7

Tilting Confidence Limits (maximum-likelihood tilting):
       2.5%  5%     95%    97.5%
mean   265   274.4  434.2  458.7
```

FIGURE 18.18 S-PLUS output for bootstrapping the mean of the Seattle real estate sales price data. From this output you can obtain the bootstrap t and percentile intervals, which are not accurate for these data. You can also obtain the BCa and tilting intervals, the recommended methods.

The bootstrap t and percentile intervals aren't reliable when the sampling distribution of the statistic is skewed. Figure 18.18 shows software output that allows us to obtain more accurate confidence intervals. The BCa interval is

$$(329.3 - 59.2, 329.3 + 126.4) = (270.0, 455.7)$$

and the tilting interval is

$$(329.3 - 64.3, 329.3 + 129.5) = (265.0, 458.7)$$

The intervals agree closely (we usually find only small differences between highly accurate procedures). Both are strongly asymmetrical—the upper endpoint is about twice as far from the sample mean as the lower endpoint—reflecting the strong right-skewness of the data.

In this example, both endpoints of the less-accurate procedures—t, bootstrap t, and percentile intervals—are too low. These intervals are too likely (greater than 2.5%) to fall below the population mean and are not likely enough to fall above the population mean. They give a biased picture of where the true mean is likely to be. If you use these intervals to budget how much you would need to be 95% confident of affording an average home, your estimate would be too low.

APPLY YOUR KNOWLEDGE

18.33 Comparing intervals. Use the software output in Figure 18.18 to give the bootstrap t and percentile 95% confidence intervals for the mean μ of all 2002 real estate sales in Seattle. Also give the traditional one-sample t interval, $\bar{x} \pm t^*s/\sqrt{n}$. Example 18.11 reports the BCa and tilting intervals. Make a picture that compares all five confidence intervals by drawing a vertical line at $\bar{x}$ and placing the intervals one above the other on this line. Describe how the intervals compare. In practical terms, what kind of errors would you make by using a *t* interval or percentile interval instead of a tilting or BCa interval?

18.34 Comparing intervals. The bootstrap distribution of the 25% trimmed mean for the Seattle real estate sales (Figure 18.8) is not strongly skewed. We were willing in Example 18.6 to give the 95% bootstrap *t* confidence interval for the trimmed mean of the population. Was that wise? Bootstrap the trimmed mean and give all of the bootstrap 95% confidence intervals: *t*, percentile, BCa, and tilting. Make a picture that compares these intervals by drawing a vertical line at $\bar{x}_{25\%}$ and placing the intervals one above the other on this line. Describe how the intervals compare. Is the *t* interval reasonably accurate?

18.35 Wages and length of service. Table 10.1 (page 587) reports the wages and length of service for a random sample of 59 women who hold customer service jobs in Indiana banks. Exercise 18.32 asked you to give a bootstrap confidence interval for the population correlation between these variables. In practice, you would use the BCa or tilting method. Bootstrap the correlation from the sample in Table 10.1 and compare the BCa and tilting intervals with the bootstrap *t* and percentile intervals. If you did Exercise 18.32, explain why the *t* and percentile intervals you now obtain differ slightly from those you found in the earlier exercise.

How the BCa and tilting intervals work

The BCa confidence interval endpoints are percentiles of the bootstrap distribution that are adjusted to correct for bias and skewness in the distribution. For example, the endpoints of the BCa 95% confidence interval for the mean of the 2002 Seattle real estate data are the 4.3th and 98.8th percentiles of the bootstrap distribution, rather than 2.5th and 97.5th percentiles. If the statistic is biased upward (that is, if it tends to be too large), the BCa bias correction moves the endpoints to the left. If the bootstrap distribution is skewed to the right, the BCa incorporates a correction to move the endpoints even farther to the right; this may seem counterintuitive, but it is the correct action. Details of the computations are a bit advanced, so we rely on software to calculate these intervals.

The tilting interval, in contrast, works by adjusting the process of randomly forming resamples. To calculate the left endpoint of the interval, it starts by finding a pseudopopulation that is similar to the sample except that the bootstrap distribution from this population has its 97.5th percentile equal to the observed statistic from our SRS. Then the left endpoint of the tilting interval is the parameter of that pseudo-population. Similarly, the right endpoint of the interval is the parameter of a pseudo-population whose bootstrap distribution has its 2.5th percentile equal to the observed statistic of the SRS. We again rely on software to handle the calculations.

Bootstrap tilting is more efficient than other bootstrap intervals, requiring only about 1/37 as many resamples as BCa intervals for similarly accurate results. If we require high accuracy, 1000 resamples is often not enough for the BCa interval; 5000 resamples would be better.

Section 18.5 Summary

- Both bootstrap t and (when they exist) traditional z and t confidence intervals require statistics with small bias and sampling distributions close to Normal. We can check these conditions by examining the bootstrap distribution for bias and lack of Normality.

- The **bootstrap percentile confidence interval** for 95% confidence is the interval from the 2.5th percentile to the 97.5th percentile of the bootstrap distribution. Agreement between the bootstrap t and percentile intervals is an added check on the conditions needed by the t interval. Do not use t or percentile intervals if these conditions are not met.

- When bias or skewness is present in the bootstrap distribution, use either a **bootstrap tilting** or **BCa interval**. The t and percentile intervals are inaccurate under these circumstances unless the sample sizes are very large. The tilting and BCa confidence intervals adjust for bias and skewness and are generally accurate except for small samples.

Section 18.5 Exercises

18.36 CLEC repair times. The CLEC data of Example 18.7 are strongly skewed to the right. The 23 CLEC repair times (in hours) are

26.62	8.60	0	21.15	8.33	20.28	96.32	17.97
3.42	0.07	24.38	19.88	14.33	5.45	5.40	2.68
0	24.20	22.13	18.57	20.00	14.13	5.80	

(a) Make a histogram and Normal quantile plot of the sample data, and find the sample mean.

(b) Bootstrap the mean of the data. Plot the bootstrap distribution. Is it Normal? Do you expect any of the confidence intervals to be inaccurate? Why or why not?

(c) Find the bootstrap standard error and use it to create a 95% t confidence interval.

(d) Find the 95% percentile, BCa, and tilting intervals.

(e) How do the intervals compare? Briefly explain the reasons for any differences.

(f) Suppose you were using these data and confidence intervals to determine staffing levels for the coming year that you are confident would match the demand. What kind of errors would you make by using a t interval or percentile interval instead of a tilting or BCa interval?

18.5 Bootstrap Confidence Intervals

18.37 Mean difference in repair times. In Example 18.7 we looked at the mean difference between repair times for Verizon (ILEC) customers and customers of competing carriers (CLECs). The bootstrap distribution was non-Normal with strong right-skewness, making a t confidence interval inappropriate.

(a) Bootstrap the difference in means for the repair time data.

(b) Find the BCa and bootstrap tilting 95% confidence intervals. Do they agree closely? What do you conclude about mean repair times for all customers?

(c) In practical terms, what kind of errors would you make by using a t interval or percentile interval instead of a tilting or BCa interval?

18.38 Really Normal data. In Exercise 18.19 you bootstrapped the mean of a simulated SRS from the standard Normal distribution $N(0, 1)$ and found the standard error for the mean.

(a) Create the 95% bootstrap percentile confidence interval for the mean of the population. We know that the population mean is in fact 0. Does the confidence interval capture this mean?

(b) Compare the bootstrap percentile and bootstrap t intervals. Do these agree closely enough to indicate that these intervals are accurate?

18.39 Clothing for runners. In Exercise 18.21 you found the bootstrap standard error of the standard deviation of the weights of male runners. Your company is also interested in the average weight of its customers.

(a) Give the 95% t confidence interval for the mean weight of runners using the standard error $s/\sqrt{n}$ computed by formula.

(b) Are there any data points that might strongly influence this confidence interval?

(c) Give a 95% bootstrap percentile confidence interval for the mean. Compare your interval with your work in (a).

(d) What conclusions can you draw about the population?

18.40 Earnings of black male bank workers. Table 1.8 (page 31) gives the earnings for a random sample of black male hourly workers at National Bank.

(a) Make a histogram and Normal quantile plot of the data. Choose a statistic to measure the center of the distribution. Justify your choice in terms of the shape of the distribution and the size of the sample.

(b) Bootstrap your statistic and report its standard error.

(c) Choose a confidence interval based on the shape and bias of the bootstrap distribution, and calculate it. What do you conclude about the typical salary of black male hourly workers at National Bank?

18.41 Bootstrap to check traditional inference. Bootstrapping is a good way to check whether traditional inference methods are accurate for a given sample. Consider the following data:

109	123	118	99	121	134	126	114	129	123	171	124	111	125	128
154	121	123	118	106	108	112	103	125	137	121	102	135	109	115
125	132	134	126	116	105	133	111	112	118	117	105	107		

(a) Examine the data graphically. Do they appear to violate any of the conditions needed to use the one-sample t confidence interval for the population mean?

(b) Calculate the 95% one-sample t confidence interval for this sample.
(c) Bootstrap the mean, and inspect the bootstrap distribution. Does it suggest that a t interval should be reasonably accurate?
(d) Find the 95% bootstrap percentile interval. Does it agree with the one-sample t interval? What do you conclude about the accuracy of the one-sample t interval here?

18.42 **More on checking traditional inference.** Continue to work with the data given in the previous exercise.
(a) Find the bootstrap BCa or tilting 95% confidence interval. We believe that either interval is quite accurate.
(b) Does your opinion of the robustness of the t confidence interval change when you compare it with the BCa or tilting interval?
(c) To check the accuracy of the one-sample t confidence interval, would you generally use the bootstrap percentile or BCa (or tilting) interval?

18.43 **Iowa housing prices.** Table 2.13 (page 165) gives the selling price, square footage, and age for a sample of 50 houses sold in Ames, Iowa.
(a) Make a histogram and Normal quantile plot of the prices. Based on these plots, decide which statistic—mean, trimmed mean, or median—would be the most useful measure of the price of typical houses sold in Ames.
(b) Bootstrap that statistic and find its standard error.
(c) Plot the bootstrap distribution and describe its shape and bias. Choose an appropriate 95% confidence interval for this sampling distribution, and calculate it. Why did you choose this type of interval?
(d) What conclusion do you draw about Ames houses?

18.44 **Iowa housing prices.** Bootstrap the correlation between selling price and square footage in the Ames, Iowa, housing data from Table 2.13 (page 165). Describe the bootstrap distribution, and give a 95% confidence interval that is appropriate for these data. Explain your choice of interval. State your conclusions from your analysis.

18.45 **Weight as a predictor of car mileage.** Table 18.3 gives weight in pounds and gas mileage in miles per gallon for a sample of cars from the 1990 model year.[12]
(a) Make a scatterplot of the data. Characterize the relationship. Calculate the sample correlation between weight and mileage.
(b) Bootstrap the correlation. Report an accurate confidence interval for the correlation and tell what it means.
(c) Calculate the least-squares regression line to predict mileage from weight. What is the traditional t confidence interval (page 596) for the slope of the population regression line?
(d) Bootstrap the regression model. Give a 95% percentile confidence interval for the regression slope using the bootstrap.

18.46 **Baseball salaries.** Table 18.2 gives data on a sample of 50 baseball players.
(a) Find the least-squares regression line for predicting batting average from salary.

18.6 Significance Testing Using Permutation Tests

TABLE 18.3 Weight and gas mileage of 1990 model automobiles

Weight	Mileage	Weight	Mileage	Weight	Mileage
2560	33	2840	26	3450	22
2345	33	2485	28	3145	22
1845	37	2670	27	3190	22
2260	32	2640	23	3610	23
2440	32	2655	26	2885	23
2285	26	3065	25	3480	21
2275	33	2750	24	3200	22
2350	28	2920	26	2765	21
2295	25	2780	24	3220	21
1900	34	2745	25	3480	23
2390	29	3110	21	3325	23
2075	35	2920	21	3855	18
2330	26	2645	23	3850	20
3320	20	2575	24	3195	18
2885	27	2935	23	3735	18
3310	19	2920	27	3665	18
2695	30	2985	23	3735	19
2170	33	3265	20	3415	20
2710	27	2880	21	3185	20
2775	24	2975	22	3690	19

(b) Bootstrap the regression line, and give a 95% confidence interval for the slope of the population regression line.

(c) In Example 18.10 we found bootstrap confidence intervals for the correlation between salary and batting average. Does your interval for the slope of the population line agree with the conclusion of that example that there may be no relation between salary and batting average? Explain.

18.47 The influence of outliers. We know that outliers can strongly influence statistics such as the mean and the least-squares line. The black female hourly worker data in Table 1.8 (page 31) contain a low outlier.

(a) Bootstrap the mean with and without the outlier. How does the outlier influence the shape and bias of the bootstrap distribution?

(b) Find 95% BCa intervals for the population mean from both bootstrap distributions. Discuss the differences.

18.6 Significance Testing Using Permutation Tests

We use significance tests to determine whether an observed effect, such as a difference between two means or the correlation between two variables, could reasonably be ascribed to the randomness introduced in selecting the sample. If not, we have evidence that the effect observed in the sample reflects an effect that is present in the population. The reasoning of tests goes like this:

1. Choose a statistic that measures the effect we are looking for.

2. Construct the sampling distribution that this statistic would have if the effect were *not* present in the population.

3. Locate the observed statistic on this distribution. A value in the main body of the distribution could easily occur just by chance. A value in the tail would rarely occur by chance, and so is evidence that something other than chance is operating.

null hypothesis

P-value

The statement that the effect we seek is *not* present in the population is the **null hypothesis**, H_0. The probability, calculated taking the null hypothesis to be true, that we would observe a statistic value as extreme or more extreme than the one we did observe is the **P-value**. Figure 18.19 illustrates the idea of a *P*-value. Small *P*-values are evidence against the null hypothesis and in favor of a real effect in the population. The reasoning of statistical tests is indirect and a bit subtle but is by now familiar. Tests based on resampling don't change this reasoning. They find *P*-values by resampling calculations rather than from formulas and so can be used in settings where traditional tests don't apply.

Because *P*-values are calculated by *assuming that the null hypothesis is true*, we cannot resample from the observed sample as we did earlier. In the absence of bias, resampling from the original sample creates a bootstrap distribution centered at the observed value of the statistic. We must create a distribution centered at the parameter value stated by the null hypothesis. We must obey this rule:

> **RESAMPLING RULE FOR SIGNIFICANCE TESTS**
>
> Resample in a manner that is consistent with the null hypothesis.

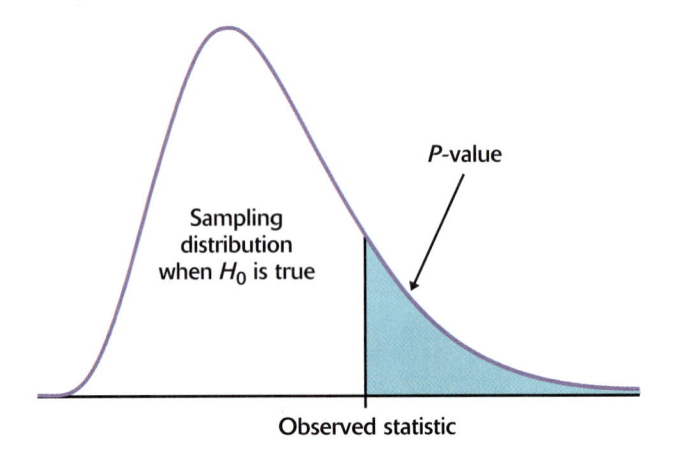

FIGURE 18.19 The *P*-value of a statistical test is found from the sampling distribution the statistic would have if the null hypothesis were true. It is the probability of a result at least as extreme as the value we actually observed.

18.6 Significance Testing Using Permutation Tests

EXAMPLE 18.12

Do reading activities increase DRP scores?

In Example 7.11 (page 464) we did a *t* test to determine whether new "directed reading activities" improved the reading ability of elementary school students, as measured by their Degree of Reading Power (DRP) score. The study assigned students at random to either the new method (treatment group, 21 students) or traditional teaching methods (control group, 23 students). Their DRP scores at the end of the study appear in Table 18.4. The statistic that measures the success of the new method is the difference in mean DRP scores,

$$\bar{x}_{\text{treatment}} - \bar{x}_{\text{control}}$$

The null hypothesis is "no difference" between the two methods. If this H_0 is true, the DRP scores in Table 18.4 do not depend on the teaching method. Each student has a DRP score that describes that child and is the same no matter which group the child is assigned to. The observed difference in group means just reflects the accident of random assignment to the two groups. Now we can see how to resample in a way that is consistent with the null hypothesis: imitate many repetitions of the random assignment, with each student always keeping his or her DRP score unchanged. Because resampling in this way scrambles the assignment of students to groups, tests based on resampling are called **permutation tests**, from the mathematical name for scrambling a group of things.

permutation test

Here is an outline of the permutation test procedure for comparing the mean DRP scores in Example 18.12:

- Choose 21 of the 44 students at random to be the treatment group; the other 23 are the control group. This is an ordinary SRS, chosen *without replacement*. It is called a **permutation resample.** Calculate the mean DRP score in each group, using the individual DRP scores in Table 18.4. The difference between these means is our statistic.

permutation resample

- Repeat this resampling from the 44 students hundreds of times. The distribution of the statistic from these resamples forms the sampling distribution under the condition that H_0 is true. It is called a **permutation distribution.**

permutation distribution

TABLE 18.4 DRP scores for third-graders

Treatment group				Control group			
24	61	59	46	42	33	46	37
43	44	52	43	43	41	10	42
58	67	62	57	55	19	17	55
71	49	54		26	54	60	28
43	53	57		62	20	53	48
49	56	33		37	85	42	

```
                    ┌─────────────────────────┐
                    │   24, 61 | 42, 33, 46, 37  │
                    │ x̄₁ − x̄₂ = 42.5 − 39.5 = 3.0│
                    └─────────────────────────┘
```

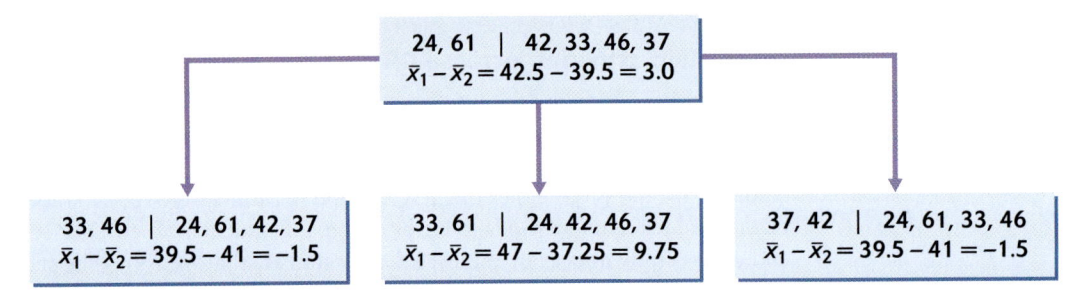

33, 46 | 24, 61, 42, 37	33, 61 | 24, 42, 46, 37	37, 42 | 24, 61, 33, 46
x̄₁ − x̄₂ = 39.5 − 41 = −1.5	x̄₁ − x̄₂ = 47 − 37.25 = 9.75	x̄₁ − x̄₂ = 39.5 − 41 = −1.5

FIGURE 18.20 The idea of permutation resampling. The top box shows the outcomes of a study with four subjects in one group and two in the other. The boxes below show three permutation resamples. The values of the statistic for many such resamples form the permutation distribution.

- The value of the statistic actually observed in the study was

$$\bar{x}_{\text{treatment}} - \bar{x}_{\text{control}} = 51.476 - 41.522 = 9.954$$

Locate this value on the permutation distribution to get the *P*-value.

Figure 18.20 illustrates permutation resampling on a small scale. The top box shows the results of a study with 4 subjects in the treatment group and 2 subjects in the control group. A permutation resample chooses an SRS of 4 of the 6 subjects to form the treatment group. The remaining 2 are the control group. The results of three permutation resamples appear below the original results, along with the statistic (difference in group means) for each.

EXAMPLE 18.13 **The permutation test for DRP scores**

Figure 18.21 shows the permutation distribution of the difference in means based on 999 permutation resamples from the DRP data in Table 18.4. The solid line in the figure marks the value of the statistic for the original sample, 9.954.

We seek evidence that the treatment increases DRP scores, so the hypotheses are

$$H_0: \mu_{\text{treatment}} - \mu_{\text{control}} = 0$$
$$H_a: \mu_{\text{treatment}} - \mu_{\text{control}} > 0$$

The *P*-value for the one-sided test is the probability that the difference in means is 9.954 or greater, calculated taking the null hypothesis to be true. The permutation distribution in Figure 18.21 shows how the statistic would vary if the null hypothesis were true. So the proportion of observations greater than 9.954 estimates the *P*-value. A look at the resampling results finds that 14 of the 999 resamples gave a value of 9.954 or larger.

The proportion of samples that exceed the observed value 9.954 is 14/999, or 0.014. Here is a last refinement. Recall from Chapter 8 that we can improve the estimate of a population proportion by adding two successes and two failures to the sample. It turns out that we can similarly improve the estimate of the *P*-value

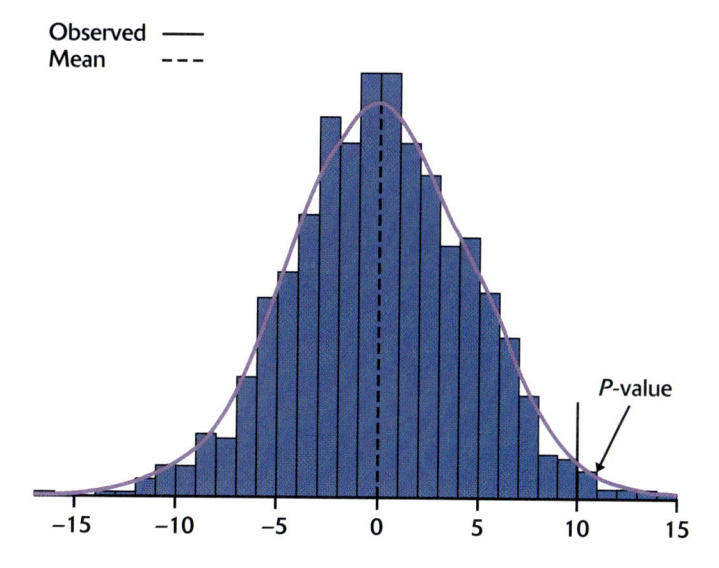

FIGURE 18.21 The permutation distribution of the statistic $\bar{x}_{treatment} - \bar{x}_{control}$ based on the DRP scores of 44 students. The observed difference in means, 9.954, is in the right tail.

by adding one sample result above the observed statistic. The final permutation test estimate of the P-value is

$$\frac{14+1}{999+1} = \frac{15}{1000} = 0.015$$

The data give good evidence that the new method beats the standard method.

Figure 18.21 shows that the permutation distribution has a roughly Normal shape. Because the permutation distribution approximates the sampling distribution, we now know that the sampling distribution is close to Normal. When the sampling distribution is close to Normal, we can use the usual two-sample t test. Example 7.11 shows that the t test gives $P = 0.013$, very close to the P-value from the permutation test.

Using software

In principle, you can program almost any statistical software to do a permutation test. It is much more convenient to use software that automates the process of resampling, calculating the statistic, forming the resampling distribution, and finding the P-value.

The commands that do this in S-PLUS are

```
permDRP = permutationTestMeans
   (data = DRP, treatment = group, alternative="greater")
plot(permDRP)
permDRP
```

The first command uses the "group" variable from the DRP data set to determine groups, calculates the difference in means for each remaining variable (in this case, only "score"), creates the permutation distribution, and calculates the one-sided P-value for the specified alternative hypothesis (if the alternative is omitted, then two-sided P-values are computed). The `plot` command produces the permutation distribution in Figure 18.21 from the DRP data. The final command prints this summary of results:

```
Number of Replications: 999

Summary Statistics:
       Observed    Mean      SE  alternative  p.value
score     9.954  0.07153  4.421      greater    0.015
```

The output makes it clear, by giving "greater" as the alternative hypothesis, that 0.015 is the one-sided P-value. For a two-sided test, double the one-sided P-value to get $P = 0.030$.

APPLY YOUR KNOWLEDGE

18.48 **Permutation test by hand.** To illustrate the process, let's perform a permutation test for a small random subset of the DRP data. Here are the data:

Treatment group	24	61			
Control group	42	33	46	37	

(a) Calculate the difference in means $\bar{x}_{treatment} - \bar{x}_{control}$ between the two groups. This is the observed value of the statistic.

(b) Resample: Start with the 6 scores and choose an SRS of 2 scores to form the treatment group for the first resample. You can do this by labeling the scores 1 to 6 and using consecutive random digits from Table B, or by rolling a die to choose from 1 to 6 at random. Using either method, be sure to skip repeated digits. A resample is an ordinary SRS, without replacement. The remaining 4 scores are the control group. What is the difference in group means for this resample?

(c) Repeat step (b) 20 times to get 20 resamples and 20 values of the statistic. Make a histogram of the distribution of these 20 values. This is the permutation distribution for your resamples.

(d) What proportion of the 20 statistic values were equal to or greater than the original value in part (a)? You have just estimated the one-sided P-value for the original 6 observations.

18.49 **Have Seattle real estate prices increased?** Table 18.1 contains the selling prices for a random sample of 50 Seattle real estate transactions in 2002. Table 18.5 contains a similar random sample of sales in 2001. Test whether the means of two random samples of the 2001 and 2002 Seattle real estate sales data are significantly different.

(a) State the null and alternative hypotheses.

(b) Perform a two-sample t test. What is the P-value?

18.6 Significance Testing Using Permutation Tests

TABLE 18.5 Selling prices (in $1000) for an SRS of 50 Seattle real estate sales in 2001

419	55.268	65	210	510.728	212.2	152.720	266.6	69.427	125
191	451	469	310	325	50	675	140	105.5	285
320	305	255	95.179	346	199	450	280	205.5	135
190	452.5	335	455	291.905	239.9	369.95	569	481	475
495	195	237.5	143	218.95	239	710	172	228.5	270

(c) Perform a permutation test on the difference in means. What is the P-value? Compare it with the P-value you found in part (b). What do you conclude based on the tests?

(d) Find a bootstrap tilting or BCa 95% confidence interval for the difference in means. How is the interval related to your conclusion in (c)?

18.50 Assets to liabilities ratio. Case 7.2 (page 476) compared the ratio of current assets to current liabilities for samples of 68 healthy firms and 33 failed firms. We conjecture that the mean ratio is higher in the population of healthy firms than among failed firms.

(a) State null and alternative hypotheses.

(b) Perform a two-sample t test using the data in Table 7.4. What is the P-value?

(c) Perform a permutation test on the difference in means. What is the P-value? Compare it with the P-value found in part (b). What do you conclude based on the tests?

(d) Find a bootstrap tilting or BCa 95% confidence interval for the difference in means. How is the interval related to the test result in (c)?

Permutation tests in practice

Advantages of Permutation Tests

In Example 18.13, the permutation test and the two-sample t test gave very similar P-values. Permutation tests have these advantages over t tests:

- The t test gives accurate P-values if the sampling distribution of the difference in means is at least roughly Normal. The permutation test gives accurate P-values even when the sampling distribution is not close to Normal.

- We can directly check the Normality of the sampling distribution by looking at the permutation distribution.

Permutation tests provide a "gold standard" for assessing two-sample t tests. If the two P-values differ considerably, it usually indicates that the conditions for the two-sample t don't hold for these data. Because permutation tests give accurate P-values even when the sampling distribution is skewed, they are often used when accuracy is very important. Here is an example.

EXAMPLE 18.14

CASE 18.1

Telecommunications data: test of difference in means

In Example 18.7, we looked at the difference in means between repair times for 1664 Verizon (ILEC) customers and 23 customers of competing companies (CLECs). Figure 18.9 shows both distributions. Penalties are assessed if a significance test concludes at the 1% significance level that CLEC customers are receiving inferior service. A one-sided test is used, because the alternative of interest to the public utilities commission (PUC) is that CLEC customers are disadvantaged.

Because the distributions are strongly skewed and the sample sizes are very different, two-sample t tests are inaccurate. An inaccurate testing procedure might declare 3% of tests significant at the 1% level when in fact the two groups of customers are treated identically, so that only 1% of tests should in the long run be significant. Errors like this would cost Verizon substantial sums of money.

Verizon uses permutation tests with 500,000 resamples for high accuracy, using custom software based on S-PLUS. Depending on the preferences of the state PUC, one of three statistics is chosen: the difference in means, $\bar{x}_1 - \bar{x}_2$; the pooled-variance t statistic, or a modified t statistic in which only the standard deviation of the larger group is used to determine the standard error. The last statistic prevents the large variation in the small group from inflating the standard error.

To perform the permutation test, we randomly redistribute the repair times into two groups that are the same sizes as the two original samples. Each repair time appears once in the data in each resample, but some repair times from the ILEC group move to CLEC, and vice versa. We calculate the test statistics for each resample and create the permutation distribution for each test statistic. The P-values are the proportions of the resamples with statistics that exceed the observed statistics.

Here are the P-values for the three tests on the Verizon data, using 500,000 permutations. The corresponding t test P-values, obtained by comparing the t statistic with t critical values, are shown for comparison.

Test statistic	t test P-value	Permutation test P-value
$\bar{x}_1 - \bar{x}_2$		0.0183
Pooled t statistic	0.0045	0.0183
Modified t statistic	0.0044	0.0195

The t test results are off by a factor of more than 4 because they do not take skewness into account. The t test suggests that the differences are significant at the 1% level, but the more accurate P-values from the permutation test indicate otherwise. Figure 18.22 shows the permutation distribution of the first statistic, the difference in sample means. The strong skewness implies that t tests will be inaccurate.

Other data sets Verizon encounters are similar to this one in being strongly skewed with imbalanced sample sizes. If Verizon and the PUCs used t tests instead of the more accurate permutation tests, there would be about four times too many false-positives (cases where a significance test indicates

18.6 Significance Testing Using Permutation Tests

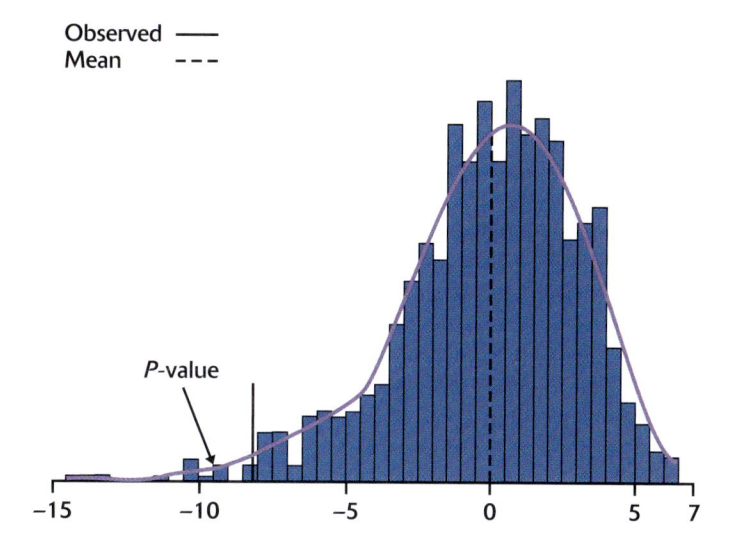

FIGURE 18.22 The permutation distribution of the difference of means $\bar{x}_1 - \bar{x}_2$ for the Verizon repair time data. The distribution is skewed left. The observed difference in means, -8.098, is in the left tail.

a significant difference even though the corresponding populations are the same), which would result in substantial financial penalties.

Data from an Entire Population

A subtle difference between confidence intervals and significance tests is that confidence intervals require the distinction between sample and population but tests do not. If we have data on an entire population—say, all employees of a large corporation—we don't need a confidence interval to estimate the difference between the mean salaries of male and female employees. We can calculate the means for all men and for all women and get an exact answer. But it still makes sense to ask, "Is the difference in means so large that it would rarely occur just by chance?" A test and its *P*-value answer that question.

Permutation tests are a convenient way to answer such questions. In carrying out the test we pay no attention to whether the data are a sample or an entire population. The resampling assigns the full set of observed salaries at random to men and women and builds a permutation distribution from repeated random assignments. We can then see if the observed difference in mean salaries is so large that it would rarely occur if gender did not matter.

When Are Permutation Tests Valid?

The two-sample *t* test starts from the condition that the sampling distribution of $\bar{x}_1 - \bar{x}_2$ is Normal. This is the case if both populations have Normal distributions, and it is approximately true for large samples from non-Normal populations because of the central limit theorem. The central limit theorem helps explain the robustness of the two-sample *t* test. The two-sample *t* test works well when both populations are symmetric, or when the populations have mild skewness in the same direction and the two sample sizes are similar.

The permutation test completely removes the Normality condition. The tradeoff is that it requires the two populations to have identical distributions when the null hypothesis is true—not only the same means, but also the same spreads and shapes. It needs this to be able to move observations randomly between groups. In practice it is robust against different distributions, except for different spreads when the sample sizes are not similar. Our preferred version of the two-sample t allows different standard deviations in the two populations.

However, this is rarely a reason to choose the t test over the permutation test, for two reasons. First, even if you notice that the two samples have different standard deviations, this does not necessarily mean that the population standard deviations differ. Particularly for skewed populations, the sample standard deviations may be very different even when the population standard deviations are the same. Second, it is usually reasonable to assume that the distributions are approximately the same if the null hypothesis is true. In practice robustness against unequal standard deviations is less important for hypothesis testing than for confidence intervals.

In Example 18.14, the distributions are strongly skewed, ruling out the t test. The permutation test is valid if the repair time distributions for Verizon customers and CLEC customers are the same if the null hypothesis is true—in other words, that all customers are treated the same.

Sources of Variation

Just as in the case of bootstrap confidence intervals, permutation tests are subject to two sources of random variability: the original sample is chosen at random from the population, and the resamples are chosen at random from the sample. Again as in the case of the bootstrap, the added variation due to resampling is usually small and can be made as small as we like by increasing the number of resamples. For example, Verizon uses 500,000 resamples.

For most purposes, 999 resamples are sufficient. If stakes are high or P-values are near a critical value (for example, near 0.01 in the Verizon example), increase the number of resamples. Here is a helpful guideline: If the true (one-sided) P-value is p, the standard deviation of the estimated P-value is approximately $\sqrt{p(1-p)/B}$, where B is the number of resamples. You can choose B to obtain a desired level of accuracy.

APPLY YOUR KNOWLEDGE

18.51 **Choosing the number of resamples.** The estimated P-value for the DRP study (Example 18.13) based on 999 resamples is 0.015. For the Verizon study (Example 18.14) the estimated P-value for the test based on $\bar{x}_1 - \bar{x}_2$ is 0.0183 based on 500,000 resamples. What is the approximate standard deviation of each of these estimated P-values? (Use each P as an estimate of the unknown true P-value p.)

18.52 **Validity of test methods.** You want to test the equality of the means of two populations. Sketch density curves for two populations for which

(a) a permutation test is valid but a t test is not.

(b) both permutation and t tests are valid.

(c) a t test is valid but a permutation test is not.

18.6 Significance Testing Using Permutation Tests

Permutation tests in other settings

The bootstrap procedure can replace many different formula-based confidence intervals, provided that we resample in a way that matches the setting. The permutation test is also a general method that can be adapted to various settings.

> **GENERAL PROCEDURE FOR PERMUTATION TESTS**
>
> To carry out a permutation test based on a statistic that measures the size of an effect of interest:
>
> 1. Compute the statistic for the original data.
>
> 2. Choose permutation resamples from the data without replacement in a way that is consistent with the null hypothesis of the test and with the study design. Construct the permutation distribution of the statistic from its values in a large number of resamples.
>
> 3. Find the *P*-value by locating the original statistic on the permutation distribution.

Formula methods generally obtain *P*-values from a standard distribution such as *t* or *F*, using either tables or software algorithms. The test statistic must be standardized so that it has the required distribution when the null hypothesis is true. This is why the two-sample *t* test uses $t = (\bar{x}_1 - \bar{x}_2)/\sqrt{(s_1^2/n_1) + (s_2^2/n_2)}$ rather than the simpler $\bar{x}_1 - \bar{x}_2$. Permutation tests, in contrast, generate a sampling distribution on the fly from the data and the chosen statistic. This allows greater flexibility in the choice of statistic in Step 1 of the procedure.

Permutation Test for Matched Pairs

The key step in the general procedure for permutation tests is to form permutation resamples in a way that is consistent with the study design and with the null hypothesis. Our examples and exercises to this point have concerned two-sample settings. How must we modify our procedure for a matched pairs design?

EXAMPLE 18.15 **Effects of language instruction**

Example 7.7 (page 443) looked at scores of 20 executives on a French language listening test taken both before and after a language course. The "before" and "after" data are not two independent samples, because each executive's scores reflect his or her previous knowledge of French and other individual factors. The scores appear in Table 7.2. How shall we carry out a permutation test?

The null hypothesis says that the language course has no effect on test scores. If this is true, each executive's before and after scores are just two measurements of that person's understanding of French. The "before" and "after" have no meaning because the course has no effect. Resampling randomly assigns one of each executive's two scores to "before" and the other to "after." We do not mix scores from different people because that isn't consistent with the pairing in the study design.

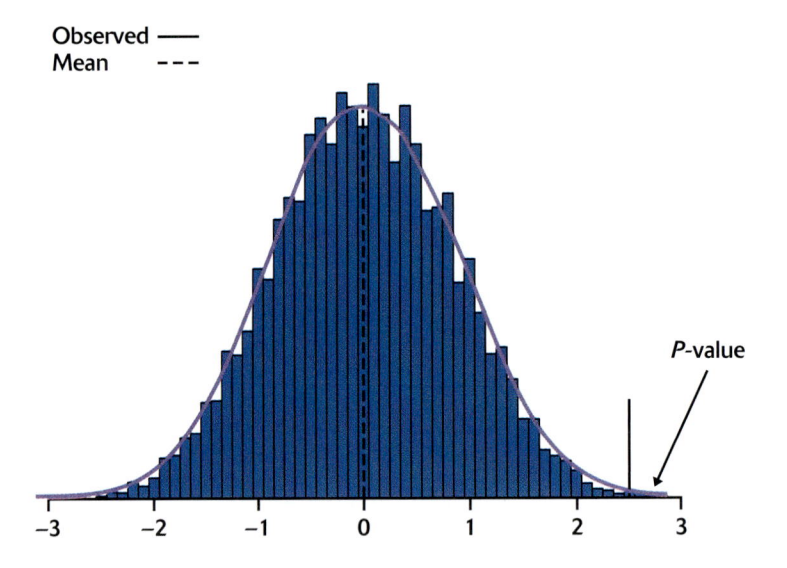

FIGURE 18.23 The permutation distribution for the mean difference (score after instruction minus score before instruction) from 9999 paired resamples from the data in Table 7.2. The observed difference in means, 2.5, is in the right tail.

After forming the "before" and "after" scores by randomly permuting each matched pair separately, calculate the difference (after − before) and the mean difference for the 20 pairs of scores. This statistic measures the effect of the course. Figure 18.23 shows the permutation distribution for 9999 resamples from the data in Table 7.2. The observed difference is far out in the right tail, P-value = 0.0015. There is very strong evidence that the course increases French listening ability.

Permutation Test for the Significance of a Relationship

Permutation testing can be used to test the significance of a relationship between two variables. For example, in Case 18.3 we looked at the relationship between baseball players' batting averages and salaries.

The null hypothesis is that there is no relationship. In that case, salaries are assigned to players for reasons that have nothing to do with batting averages. We can resample in a way consistent with the null hypothesis by permuting the observed salaries among the players at random.

Take the correlation as the test statistic. For every resample, calculate the correlation between the batting averages (in their original order) and salaries (in the reshuffled order). The P-value is the proportion of the resamples with correlation larger than the original correlation.

When Can We Use Permutation Tests?

We can use a permutation test only when we can see how to resample in a way that is consistent with the study design and with the null hypothesis. We now know how to do this for the following types of problems:

- **Two-sample problems** when the null hypothesis says that the two populations are identical. We may wish to compare population means, proportions, standard deviations, or other statistics. You may recall from Section 7.3 that traditional tests for comparing population standard deviations work very poorly. Permutation tests help satisfy this need.

- **Matched pairs designs** when the null hypothesis says that there are only random differences within pairs. A variety of comparisons is again possible.

- **Relationships between two quantitative variables** when the null hypothesis says that the variables are not related. The correlation is the most common measure, but not the only one.

These settings share the characteristic that the null hypothesis specifies a simple situation such as two identical populations or two unrelated variables. We can see how to resample in a way that matches these situations. Permutation tests can't be used for testing hypotheses about a single population, comparing populations that differ even under the null hypothesis, or testing general relationships. In these settings, we don't know how to resample in a way that matches the null hypothesis. Researchers are developing resampling methods for these and other settings, so stay tuned.

When we can't do a permutation test, we can often calculate a bootstrap confidence interval instead. If the confidence interval fails to include the null hypothesis value, then we reject H_0 at the corresponding significance level. This is not as accurate as doing a permutation test, but a confidence interval estimates the size of an effect as well as giving some information about its statistical significance. Even when a test is possible, it is often helpful to report a confidence interval along with the test result. Confidence intervals don't assume that a null hypothesis is true, so we use bootstrap resampling with replacement rather than permutation resampling without replacement.

APPLY YOUR KNOWLEDGE

18.53 **Comparing proportions: exclusive franchise territories.** Case 9.1 (page 549) looked at the relationship between the presence of an exclusive-territory clause and the survival of new franchise firms. Exclusive-territory clauses allow the local franchise outlet to be the sole representative of the franchise in a specified territory. Firms were classified as successful or not based on whether or not they were still franchising as of a certain date. Here is a summary of the findings for firms with and without an exclusive-territory clause in their contract with local franchises:

	Firms n	Successes X	Proportion $\hat{p} = X/n$
Exclusive-territory clause	142	108	0.761
No exclusive-territory clause	28	15	0.536
Total	170	123	0.7235

(a) We conjecture that exclusive-territory clauses increase the chance of success. State appropriate null and alternative hypotheses in terms of population proportions.

(b) Perform the z test (page 527) for your hypotheses.

(c) Perform a permutation test based on the difference in the sample proportions $\hat{p}_1 - \hat{p}_2$. Explain carefully how the resampling is consistent with the null hypothesis. Compare your result with part (b).

(d) Based on your permutation test P-value, what do you conclude about the effect of exclusive-territory clauses?

(e) Give a bootstrap tilting or BCa interval for the difference between the two population proportions. Explain how the interval is consistent with the permutation test.

18.54 Matched pairs: designing controls. Exercise 7.40 (page 458) describes a study in which 25 right-handed subjects were asked to turn a knob clockwise and counterclockwise (in random order). The response variable is the time needed to move an indicator a fixed distance. We conjecture that clockwise movement is easier for right-handed people.

(a) State null and alternative hypotheses in terms of mean times. Carefully identify the parameters in your hypotheses.

(b) Perform a matched pairs permutation test. What is the P-value? What do you conclude about designing controls?

(c) Graph the permutation distribution and indicate the region that corresponds to the P-value.

18.55 Correlation: salary and batting average. Table 18.2 contains the salaries and batting averages of a random sample of 50 major league baseball players. We wonder if these variables are correlated in the population of all players.

(a) State the null and alternative hypotheses.

(b) Perform a permutation test based on the sample correlation. Report the P-value and draw a conclusion.

SECTION 18.6 SUMMARY

- **Permutation tests** are significance tests based on **permutation resamples** drawn at random from the original data. Permutation resamples are drawn **without replacement**, in contrast to bootstrap samples, which are drawn with replacement.

- Permutation resamples must be drawn in a way that is consistent with the null hypothesis and with the study design. In a **two-sample design**, the null hypothesis says that the two populations are identical. Resampling randomly reassigns observations to the two groups. In a **matched pairs** design, randomly permute the two observations within each pair separately. To test the hypothesis of **no relationship** between two variables, randomly reassign values of one of the two variables.

- The **permutation distribution** of a suitable statistic is formed by the values of the statistic in a large number of resamples. Find the P-value of the test by locating the original value of the statistic on the permutation distribution.

- When they can be used, permutation tests have great advantages. They do not require specific population shapes such as Normality. They apply to a variety of statistics, not just to statistics that have a simple distribution

18.6 Significance Testing Using Permutation Tests

under the null hypothesis. They can give very accurate *P*-values, regardless of the shape and size of the population (if enough permutations are used).

- It is often useful to give a confidence interval along with a test. To create a confidence interval, we no longer assume the null hypothesis is true, so we use bootstrap resampling rather than permutation resampling.

SECTION 18.6 EXERCISES

18.56 Female basketball players. Here are heights (inches) of professional female basketball players who are centers and forwards. We wonder if the two positions differ in average height.

Forwards											
69	72	71	66	76	74	71	66	68	67	70	65	72
70	68	73	66	68	67	64	71	70	74	70	75	75
69	72	71	70	71	68	70	75	72	66	72	70	69
Centers
72	70	72	69	73	71	72	68	68	71	66	68	71
73	73	70	68	70	75	68

(a) Make a back-to-back stemplot of the data. How do the two distributions compare?

(b) State null and alternative hypotheses. Do a permutation test for the difference in means of the two groups. Give the *P*-value and draw a conclusion.

18.57 Reaction time. Table 2.12 (page 163) gives reaction times for a person playing a computer game. The data contain outliers, so we will measure center by the median or trimmed mean rather than the mean.

(a) State null and alternative hypotheses.

(b) Perform a permutation test for the difference in medians. Describe the permutation distribution.

(c) Perform a permutation test for the difference in 25% trimmed means. Examine the permutation distribution. How does it compare with the permutation distribution for the median?

(d) Draw a conclusion, using the *P*-value(s) as evidence.

18.58 One- or two-sided? A customer complains to the owner of an independent fast-food restaurant that the restaurant is discriminating against the elderly. The customer claims that people 60 years old and older are given fewer french fries than people under 60. The owner responds by gathering data, collected without the knowledge of the employees so as not to affect their behavior. Here are data on the weight of french fries (grams) for the two groups of customers:

Age < 60:	75	77	80	69	73	76	78	74	75	81
Age ≥ 60:	68	74	77	71	73	75	80	77	78	72

(a) Display the two data sets in a back-to-back stemplot. Do they appear substantially different?

(b) If we perform a permutation test using the mean for "< 60" minus the mean for "≥ 60," should the alternative hypothesis be two-sided, greater, or less? Explain.

(c) Perform a permutation test using the chosen alternative hypothesis and give the *P*-value. What should the owner report to the customer?

18.59 "No Sweat" labels. Example 8.6 (page 527) presents a significance test comparing the proportion of men and women who pay attention to a "No Sweat" label when buying a garment.

(a) State the null and alternative hypotheses.

(b) Perform a permutation test based on the difference in sample proportions.

(c) Based on the shape of the permutation distribution, why does the permutation test agree closely with the *z* test in Example 8.6?

18.60 Real estate sales prices. We would like to test the null hypothesis that the two samples of Seattle real estate sales prices in 2001 and 2002 have equal medians. Data for these years appear in Tables 18.1 and 18.5. Carry out a permutation test for the difference in medians, find the *P*-value, and explain what the *P*-value tells us.

18.61 Size and age of houses. Table 2.13 (page 165) gives data for houses sold in Ames, Iowa, in 2000. The sample correlation between square footage and age is approximately $r = -0.41$, suggesting that the newer houses were smaller than the older houses. Test the hypothesis that there is no correlation between square footage and age. What do you conclude?

18.62 Calcium and blood pressure. Does added calcium intake reduce the blood pressure of African American men? In a randomized comparative double-blind trial, 10 men were given a calcium supplement for twelve weeks and 11 others received a placebo. For each subject the researchers recorded whether or not blood pressure dropped. Here are the data:[13]

Treatment	Subjects	Successes	Proportion
Calcium	10	6	0.60
Placebo	11	4	0.36
Total	21	10	0.48

Is there evidence that calcium reduces blood pressure? Use a permutation test.

18.63 More on calcium and blood pressure. The previous exercise asks for a permutation test for the difference in proportions. Now bootstrap the difference in proportions. Use the observed difference in proportions and the bootstrap standard error to create a 95% *z* interval for the difference in population proportions.

18.64 Another Verizon data set. Verizon uses permutation testing for hundreds of comparisons, comparing ILEC and CLEC distributions in different locations, for different time periods and different measures of service quality. Here is a sample from another Verizon data set, containing repair times in hours for Verizon (ILEC) and CLEC customers.

18.6 Significance Testing Using Permutation Tests

ILEC

1	1	1	1	2	2	1	1	1	1	2	2	1	1	1	1
2	2	1	1	1	1	2	3	1	1	1	1	2	3	1	1
1	1	2	3	1	1	1	2	3	1	1	1	1	2	3	
1	1	1	1	2	3	1	1	1	2	4	1	1	1	1	
2	5	1	1	1	1	2	5	1	1	1	1	2	6	1	1
1	1	2	8	1	1	1	2	15	1	1	1	2	2		

CLEC

1	1	5	5	5	1	5	5	5	5

(a) Choose and make data displays. Describe the shapes of the samples and how they differ.

(b) Perform a t test to compare the population mean repair times. Give hypotheses, the test statistic, and the P-value.

(c) Perform a permutation test for the same hypotheses using the pooled variance t statistic. Why do the two P-values differ?

(d) What does the permutation test P-value tell you?

18.65 Comparing Verizon standard deviations. We might also wish to compare the variability of repair times for ILEC and CLEC customers. For the data in the previous exercise, the F statistic for comparing sample variances is 0.869 and the corresponding P-value is 0.67. We know that this test is very sensitive to lack of Normality.

(a) Perform a two-sided permutation test on the ratio of standard deviations. What is the P-value and what does it tell you?

(b) What does a comparison of the two P-values say about the validity of the F test for these data?

18.66 Testing equality of variances. The F test for equality of variances (Section 7.3) is unreliable because it is sensitive to non-Normality in the data sets. The permutation test does not suffer from this drawback. It is therefore possible to use a permutation test to check the equal-variances condition before using the pooled version of the two-sample t test. Example 7.18 (page 490) illustrates the F test for comparing the variability of the asset-to-liability ratio in samples of healthy firms and failed firms. Do a permutation test for this comparison.

(a) State the null and alternative hypotheses.

(b) Perform a permutation test on the F statistic (ratio of sample variances). What do you conclude?

(c) Compare the permutation test P-value to that in Example 7.18. What do you conclude about the F test for equality of variances for these data?

18.67 Executives learn Spanish. Exercise 7.42 (page 459) gives the scores of 20 executives on a test of Spanish comprehension before and after a language course. We think that the course should improve comprehension scores.

(a) State the null and alternative hypotheses.

(b) Perform a paired-sample permutation test. Give the P-value and your conclusion about the effectiveness of the course.

(c) Graph the permutation distribution and indicate the region that corresponds to the P-value.

STATISTICS IN SUMMARY

Fast and inexpensive computing power allows the use of statistical procedures that require large-scale computation. Bootstrap confidence intervals and permutation tests are based on large numbers of "resamples" drawn from the data. These resampling computations replace the formulas derived from probability theory that we use in traditional confidence intervals and tests. Resampling procedures can often be used in settings that do not meet the conditions for use of formula-based procedures. These resampling procedures are becoming ever more common in statistical practice. It is possible that in the future they will largely replace some traditional procedures. Reading this chapter should enable you to do the following:

A. BOOTSTRAP

1. Explain the bootstrap resampling idea in the context of a particular data set used to estimate a particular population parameter.
2. Use software such as S-PLUS to bootstrap a statistic of your choice from a set of data. Plot the bootstrap distribution and obtain the bootstrap standard error and the bootstrap estimate of bias.
3. Based on bootstrap software output, judge whether formula-based confidence intervals that require Normal sampling distributions can be used and whether the simple bootstrap t and percentile confidence intervals can be used.
4. Obtain from bootstrap software output any of the four types of bootstrap confidence intervals for a parameter: t, percentile, BCa, and tilting. By comparing these intervals judge which are safe to use.

B. PERMUTATION TESTS

1. Recognize the settings in which we can use permutation tests. In such a setting, explain how to choose permutation resamples that are consistent with the null hypothesis and with the design of the study.
2. Use software to obtain the permutation distribution of a test statistic of your choice in settings that allow permutation tests. Give the P-value of the test.
3. Based on permutation test software output, judge whether a traditional formula-based test can be used.

CHAPTER 18 REVIEW EXERCISES

18.68 Piano lessons. Exercise 7.34 (page 456) reports the changes in reasoning scores of 34 preschool children after six months of piano lessons. Here are the changes:

| 2 | 5 | 7 | −2 | 2 | 7 | 4 | 1 | 0 | 7 | 3 | 4 | 3 | 4 | 9 | 4 | 5 |
| 2 | 9 | 6 | 0 | 3 | 6 | −1 | 3 | 4 | 6 | 7 | −2 | 7 | −3 | 3 | 4 | 4 |

(a) Make a histogram and Normal quantile plot of the data. Is the distribution approximately Normal?

(b) Find the sample mean and its standard error using formulas.
(c) Bootstrap the mean and find the bootstrap standard error. Does the bootstrap give comparable results to theoretical methods?

18.69 Uniform distribution. Your software can generate "uniform random numbers" that have the Uniform distribution on 0 to 1. See Figure 4.5 (page 247) for the density curve. Generate a sample of 50 observations from this distribution.
(a) What is the population median? Bootstrap the sample median and describe the bootstrap distribution.
(b) What is the bootstrap standard error? Compute a bootstrap t 95% confidence interval.
(c) Find the BCa or tilting 95% confidence interval. Compare with the interval in (b). Is the bootstrap t interval reliable here?

18.70 Age of personal trainers. A fitness center employs 20 personal trainers. Here are the ages in years of the female and male personal trainers working at this center:

| Male: | 25 | 26 | 23 | 32 | 35 | 29 | 30 | 28 | 31 | 32 | 29 |
| Female: | 21 | 23 | 22 | 23 | 20 | 29 | 24 | 19 | 22 | | |

(a) Make a back-to-back stemplot. Do you think the difference in mean ages will be significant?
(b) A two-sample t test gives $P < 0.001$ for the null hypothesis that the mean age of female personal trainers is equal to the mean age of male personal trainers. Do a two-sided permutation test to check the answer.
(c) What do you conclude about using the t test? What do you conclude about the mean ages of the trainers?

18.71 Stock returns. Table 2.6 (page 130) gives annual total returns for overseas and U.S. stocks over a 30-year period.
(a) Bootstrap the correlation between overseas and U.S. stocks and describe its bootstrap distribution. What is the bootstrap standard error?
(b) Is a bootstrap t confidence interval appropriate? Why or why not?
(c) Find the 95% BCa or tilting confidence interval.

18.72 Blockbuster stock. Here are data on the price of Blockbuster stock for the month of June 2002:[14]

Date	Close	Change	Date	Close	Change
6.03	27.31	−0.19	6.17	27.36	0.41
6.04	27.49	0.18	6.18	27.02	−0.34
6.05	28.41	0.92	6.19	26.63	−0.39
6.06	28.38	−0.03	6.20	26.85	0.22
6.07	27.77	−0.61	6.21	25.97	−0.88
6.10	28.02	0.25	6.24	26.39	0.42
6.11	27.84	−0.18	6.25	25.87	−0.52
6.12	27.38	−0.46	6.26	25.59	−0.28
6.13	26.20	−1.18	6.27	26.75	1.16
6.14	26.95	0.75	6.28	26.90	0.15

(a) Compute the percent change for each trading day. The standard deviation of the daily percent change is one measure of the *volatility* of the stock. Find the sample standard deviation of the percent changes.

(b) Bootstrap the standard deviation. What is its bootstrap standard error?

(c) Find the 95% bootstrap t confidence interval for the population standard deviation.

(d) Find the 95% tilting or BCa confidence interval for the standard deviation. Compare the confidence intervals and give your conclusions about the appropriateness of the bootstrap t interval.

18.73 **Real estate sales prices.** We have compared the selling prices of Seattle real estate in 2002 (Table 18.1) and 2001 (Table 18.5). Let's compare 2001 and 2000. Here are the prices (thousands of dollars) for 20 random sales in Seattle in the year 2000:

CASE 18.2

| 333 | 126.5 | 207.5 | 199.5 | 1836 | 360 | 175 | 133 | 1100 | 203 |
| 194.5 | 140 | 280 | 475 | 185 | 390 | 242 | 276 | 359 | 163.95 |

(a) Plot both the 2000 and the 2001 data. Explain what conditions needed for a two-sample t test are violated.

(b) Perform a permutation test to find the P-value for the difference in means. What do you conclude about selling prices in 2000 versus 2001?

18.74 **Radon detectors.** Exercise 7.38 (page 457) gives the following readings for 12 home radon detectors when exposed to 105 picocuries per liter of radon:

| 9.19 | 97.8 | 111.4 | 122.3 | 105.4 | 95.0 |
| 103.8 | 99.6 | 96.6 | 119.3 | 104.8 | 101.7 |

Part (a) of Exercise 7.38 judges that a t confidence interval can be used despite the skewness of the data.

(a) Give a formula-based 95% t interval for the population mean.

(b) Find the bootstrap 95% tilting interval for the mean.

(c) Look at the bootstrap distribution. Is it approximately Normal in appearance?

(d) Do you agree that the t interval is robust enough in this case? Why or why not? Keep in mind that whether the confidence interval covers 105 is important for the study's purposes.

18.75 **Use a permutation test?** The study described in the previous exercise used a one-sample t test to see if the mean reading of all detectors of this type differs from the true value 105. Can you replace this test by a permutation test? If so, carry out the test and compare results. If not, explain why not.

18.76 **Do nurses use gloves?** Nurses in an inner-city hospital were unknowingly observed on their use of latex gloves during procedures for which glove use is recommended.[15] The nurses then attended a presentation on the importance of glove use. One month after the presentation, the same nurses were observed again. Here are the proportions of procedures for which each nurse wore gloves:

Nurse	Before	After	Nurse	Before	After
1	0.500	0.857	8	0.000	1.000
2	0.500	0.833	9	0.000	0.667
3	1.000	1.000	10	0.167	1.000
4	0.000	1.000	11	0.000	0.750
5	0.000	1.000	12	0.000	1.000
6	0.000	1.000	13	0.000	1.000
7	1.000	1.000	14	1.000	1.000

(a) Why is a one-sided alternative proper here? Why must matched pairs methods be used?

(b) Do a permutation test for the difference in means. Does the test indicate that the presentation was helpful?

18.77 **Glove use by nurses, continued.** In the previous exercise, you did a one-sided permutation test to compare means before and after an intervention. If you are mainly interested in whether or not the effect of the intervention is significant at the 5% level, an alternative approach is to give a bootstrap confidence interval for the mean difference within pairs. If zero (no difference) falls outside the interval, the result is significant. Do this and report your conclusion.

18.78 **Changes in urban unemployment.** Here are the unemployment rates (percent of the labor force) in July of 2001 and 2002 for a random sample of 19 of the 331 metropolitan areas for which the Bureau of Labor Statistics publishes data:[16]

Area	2001	2002	Area	2001	2002
1	4.7	6.0	11	2.6	2.3
2	4.1	4.0	12	5.2	5.2
3	3.9	4.1	13	2.6	2.9
4	5.0	5.3	14	3.2	3.7
5	5.0	5.6	15	4.6	5.5
6	4.3	5.2	16	3.5	4.6
7	4.4	5.6	17	4.6	5.8
8	5.6	6.9	18	4.1	5.9
9	5.3	7.2	19	5.6	7.7
10	6.3	8.7			

(a) Plot the data for each year and compare the two graphs.

(b) Do a paired t test for the difference in means, and find the P-value.

(c) Do a paired-sample permutation test, and find the P-value. Compare this with your result in part (b).

18.79 Ice cream preferences. A random sample of children who came into an ice cream shop in a certain month were asked, "Do you like chocolate ice cream?" The results were:

	Girls	Boys	Total
Yes	40	30	70
No	10	15	25
Total	50	45	95

(a) Find the proportions of girls and boys who like chocolate ice cream.

(b) Perform a permutation test on the proportions, and use the P-value to determine if there is a statistically significant difference in the proportions of girls and boys who like chocolate ice cream.

18.80 Word counts in magazine ads. Is there a difference in the readability of advertisements in magazines aimed at people with varying educational levels? Here are word counts in 6 randomly selected ads from each of 3 randomly selected magazines aimed at people with high education level and 3 magazines aimed at people with middle education level:[17]

Education level	Word count
High	205 203 229 208 146 230 215 153 205 80 208 89 49 93 46 34 39 88
Medium	191 219 205 57 105 109 82 88 39 94 206 197 68 44 203 139 72 67

(a) Make histograms and Normal quantile plots for both data sets. Do the distributions appear approximately Normal? Find the means.

(b) Bootstrap the means of both data sets and find their bootstrap standard errors.

(c) Make histograms and Normal quantile plots of the bootstrap distributions. What do the plots show?

(d) Find the 95% percentile and tilting intervals for both data sets. Do the intervals for high and medium education level overlap? What does this indicate?

(e) Bootstrap the difference in means and find a 95% percentile confidence interval. Does it contain 0? What conclusions can you draw from your confidence intervals?

18.81 More on magazine ad word counts. The researchers in the study described in the previous exercise expected higher word counts in magazines aimed at people with high education levels. Do a permutation test to see if the data support this expectation. State hypotheses, give a P-value, and state your conclusions. How do your conclusions here relate to those from the previous exercise?

18.82 Hyde Park burglaries. The following table gives the number of burglaries per month in the Hyde Park neighborhood of Chicago for a period before and after the commencement of a citizen-police program.[18]

					Before					
60	44	37	54	59	69	108	89	82	61	47
72	87	60	64	50	79	78	62	72	57	57
61	55	56	62	40	44	38	37	52	59	58
69	73	92	77	75	71	68	102			
					After					
88	44	60	56	70	91	54	60	48	35	49
44	61	68	82	71	50					

(a) Plot both sets of data. Are the distributions skewed or roughly Normal?

(b) Perform a one-sided (which side?) t test on the data. Is there statistically significant evidence of a decrease in burglaries after the program began?

(c) Perform a permutation test for the difference in means, using the same alternative hypothesis as in part (b). What is the P-value? Is there a substantial difference between this P-value and the one in part (b)? Use the shapes of the distributions to explain why or why not. What do you conclude from your tests?

(d) Now do a permutation test using the opposite one-sided alternative hypothesis. Explain what this is testing, why it is not of interest to us, and why the P-value is so large.

18.83 **Hyde Park burglaries, continued.** The previous exercise applied significance tests to the Hyde Park burglary data. We might also apply confidence intervals.

(a) Bootstrap the difference in mean monthly burglary counts. Make a histogram and a Normal quantile plot of the bootstrap distribution and describe the distribution.

(b) Find the bootstrap standard error, and use it to create a 95% bootstrap t confidence interval.

(c) Find the 95% percentile confidence interval. Compare this with the t interval. Does the comparison suggest that these intervals are accurate? How do the intervals relate to the results of the tests in the previous exercise?

Notes for Chapter 18

1. S-PLUS is a registered trademark of the Insightful Corporation.

2. G. Snow and L. C. Chihara, *S-PLUS for Moore and McCabe's Introduction to the Practice of Statistics 4th ed.*, W. H. Freeman, 2003 (ISBN 0-7167-9619-8). (This is a supplement for a different book, but can be used with this book.)

3. Verizon repair time data used with the permission of Verizon.

4. T. Bjerkedal, "Acquisition of resistance in guinea pigs infected with different doses of virulent tubercle bacilli," *American Journal of Hygiene,* 72 (1960), pp. 130–148.

5. Seattle real estate sales data provided by Stan Roe of the King County Assessor's Office.

6. The 254 winning numbers and their payoffs are the *lottery* data set in S-PLUS and are originally from the New Jersey State Lottery Commission.

7. "America's Best Small Companies," *Forbes,* November 8, 1993.

8. ACORN, "Banking on discrimination: executive summary," October 1991, in *Joint Hearings before the Committee on Banking, Finance, and Urban Affairs, House of Representatives,* 102nd Congress, 2nd Session, Serial 102-120, May 7 and May 14, 1992, pp. 236–246.

9. From the *Forbes* Web site, www.forbes.com.

10. Data provided by Darlene Gordon, Purdue University, for David S. Moore and George P. McCabe, *Introduction to the Practice of Statistics,* 4th ed., W. H. Freeman and Company, 2003.

11. From www.espn.com, July 2, 2002.

12. *Consumer Reports,* April 1990, pp. 235–288. These data are part of the *fuel.frame* data set in S-PLUS.

13. Roseann M. Lyle et al., "Blood pressure and metabolic effects of calcium supplementation in normotensive white and black men," *Journal of the American Medical Association,* 257 (1987), pp. 1772–1776.

14. Data from www.nasdaq.com.

15. L. Friedland et al., "Effect of educational program on compliance with glove use in a pediatric emergency department," *American Journal of Diseases of Childhood,* 146 (1992), pp. 1355–1358.

16. From the Web site of the Bureau of Labor Statistics, www.bls.gov.

17. F. K. Shuptrine and D. D. McVicker, "Readability levels of magazine ads," *Journal of Advertising Research,* 21, No. 5 (1981), p. 47.

18. G. V. Glass, V. L. Wilson, and J. M. Gottman, *Design and Analysis of Time Series Experiments,* Colorado Associated University Press, 1975.

SOLUTIONS TO ODD-NUMBERED EXERCISES

Chapter 18

Note: In questions in this chapter for which the answers are obtained by resampling, your answers may differ slightly from those we give because your random samples may differ from ours.

18.1 (a) and (b) The means of our 20 resamples are 6.86, 1.64, 7.00, 4.27, 4.83, 4.24, 1.03, 2.30, 7.38, 4.54, 4.21, 0.70, 1.19, 2.04, 1.45, 6.86, 10.28, 0.78, 4.35, 4.32. (c) Make a stemplot of your means. (d) Bootstrap standard error = 2.658.

18.3 (a) The histogram is right-skewed. (b) The distribution is symmetric and bell-shaped, and the quantile plot shows positive skewness; while this amount of positive skewness would not be a concern in raw data, here it occurs in a bootstrap distribution, after the central limit theorem has had a chance to work. Later in the chapter we learn ways to get more accurate confidence intervals in cases like this.

18.5 (a) The distribution is roughly bell-shaped but is less so than the bootstrap distribution in Exercise 18.3. The distribution here looks slightly right-skewed, and the points plotted on a quantile plot don't lie as close to a straight line as those in the quantile plot in Exercise 18.3. (b) The bootstrap standard error for Exercise 18.3 is 3.07, and for the data in this exercise the bootstrap standard error is 7.45. For a sample of size n, the standard error of the sample mean is $s/\sqrt{n}$, where s is the sample standard deviation. A smaller sample size results in a larger standard error.

18.7 $s = 21.7$, so $s/\sqrt{50} = 3.07$. The bootstrap standard error in Exercise 18.3 is 3.07, which agrees closely with $s/\sqrt{50} = 3.07$.

18.9 The bias is 0.448. This is small compared to the observed mean of 141.8. The bootstrap distribution of most statistics mimics the shape, spread, and bias of the actual sampling distribution. Thus, we expect that the bias encountered in using $\bar{x}$ to estimate the mean survival time for all guinea pigs that receive the same experimental treatment is also small.

18.11 (a) The 25% trimmed mean is 30.1, which is smaller than the sample mean of 34.7. If we examine the histogram of the data for the 50 shoppers we see that the data are right-skewed. The trimmed mean eliminates much of the large right tail (that is, the very large values that cause the sample mean to be large), and hence the trimmed mean is smaller than the sample mean. (b) The 95% confidence interval for the 25% trimmed mean spending in the population of all shoppers is (23.74, 36.46).

18.13 The formula-based standard error is 4.08. The bootstrap standard error in Example 18.7 is 4.052, which is close to the formula-based value.

18.15 (a) The distribution looks approximately Normal and the bias is small. Thus, it meets the conditions for a bootstrap t confidence interval. (b) The bootstrap mean is .902, and the standard error is 0.114. The bootstrap t confidence interval is (0.6903, 1.1265). (c) The two-sample t confidence interval reported on page 479 is (0.65, 1.15). This interval is wider than the bootstrap t confidence interval in part (b).

18.17 (a) The two bootstrap distributions look similar. (b) The bootstrap standard error of the mean is 45.0. In Example 18.5, the bootstrap standard error of the 25% trimmed mean is 16.83. We see that the bootstrap standard error of the mean is almost three times as large as the bootstrap standard error of the 25% trimmed mean. The bootstrap distribution for the mean has greater spread (the histogram covers a range from about 225 to 475) than the bootstrap distribution for the 25% trimmed mean (the histogram covers a range from about 200 to 300). (c) Examining the bootstrap distribution of the mean in Figure 18.7 and in part (a) of this solution, we see that the bootstrap distribution is right-skewed and hence non-Normal. We should not use the bootstrap t interval if the bootstrap distribution is not Normal.

18.19 (a) There do not appear to be any significant departures from Normality. The histogram is centered at about 0, and the spread is approximately what we would expect for the $N(0, 1)$ distribution. (b) The standard error of the bootstrap mean is 0.128. (c) The bias is small (-0.00147), and the histogram of the data looks approximately Normal, so the bootstrap distribution of the mean will also look Normal. When the bootstrap distribution is approximately Normal and the bias is small, it is safe to use the bootstrap t confidence interval. The interval is ($-0.129, 0.379$).

18.21 (a) $s = 7.71$. (b) The bootstrap standard error for s is 2.23. (c) The bootstrap standard error is a little less than one-third the value of the sample standard deviation. This suggests that the sample standard deviation is only moderately accurate as an estimate of the population standard deviation. (d) Plots show that the bootstrap distribution is not Normal. Thus, it would *not* be appropriate to give a bootstrap t interval for the population standard deviation.

18.23 (a) The plots do not show any significant departures from Normality, so there is nothing in the plots to suggest that the difference in means might be non-Normal. (b) A 95% paired t confidence interval for the difference in population means is (17.4, 25.1). The interval does not contain 0 and includes only positive values. This is evidence that the minority refusal rate is larger than the white refusal rate. (c) The bootstrap distribution looks reasonably Normal. The bias is small. Thus, a bootstrap t confidence interval is appropriate here. A 95% bootstrap t confidence interval is (17.6, 24.9). This is very close to the traditional interval that we calculated in (b).

18.25 (a) This bootstrap distribution is right-skewed. The bootstrap distribution of the Verizon repair times appears to be approximately Normal. (b) The source of the skew in the bootstrap distribution of the difference in means appears to be due to the skew in the CLEC data.

18.27 (a) Mean of the sampling distribution of $\bar{x}$ is $\mu = 8.4$. Standard deviation of the sampling distribution of $\bar{x}$ is $\sigma/\sqrt{n} = 14.7/\sqrt{n}$. (b) Use S-PLUS to make the plots. The bootstrap standard error is 2.24. (c) Use S-PLUS to make the plots. For $n = 40$, SE = 1.49. For $n = 160$, SE = 0.970.

(d) If we look at the distributions, and especially the Normal quantile plots, we see that as n increases, the bootstrap distributions look more and more Normal. We also see that the standard errors decrease roughly by a factor of 2 as n increases by a factor of 4.

18.29 For a 90% bootstrap confidence interval, we would use the 5% and 95% percentiles as the endpoints.

18.31 The 95% bootstrap t interval is $(-0.144, 0.358)$, and the bootstrap percentile interval is $(-0.128, 0.356)$. There is very close agreement between the upper endpoints of both intervals. However, the lower endpoints differ somewhat, and this may indicate skewness.

18.33 A 95% bootstrap t confidence interval is $(238.7, 415.1)$. The 95% percentile confidence interval is $(252.5, 433.2)$. A 95% traditional one-sample t confidence interval is $(239.3, 419.3)$. The BCa 95% confidence interval is $(270.0, 455.7)$, and the tilting 95% confidence interval is $(265.0, 458.7)$. Now make the plot requested. The bootstrap t and traditional intervals are centered approximately on the sample mean. The bootstrap percentile interval is shifted to the right of these two. The BCa and tilting intervals are shifted even farther to the right. The latter two better reflect the skewed nature of the data. Using a t interval or the bootstrap percentile interval, we get a biased picture of what the value of the population mean is likely to be. In particular, we would underestimate its value. Any policy decisions based on these data, such as tax rates, would reflect this underestimate.

18.35 The 95% bootstrap t confidence interval is $(0.122, 0.585)$. The 95% bootstrap percentile interval is $(0.110, 0.573)$. The 95% BCa interval is $(0.133, 0.594)$. S-PLUS gives two possible tilting intervals. The 95% exponential tilting interval is $(0.1163, 0.5542)$ and the 95% maximum-likelihood tilting interval is $(0.118, 0.555)$. The BCa and tilting intervals have larger lower endpoints and are narrower than the bootstrap t and percentile intervals. If you did Exercise 18.32, the bootstrap t and percentile intervals here will differ slightly from those in Exercise 18.32 because they are based on a different bootstrap sample.

18.37 (a) and (b) The 95% BCa confidence interval is $(-18.89, -2.48)$. This interval does not include 0, so we would conclude that the mean repair times for all Verizon customers are lower than the mean repair times for all CLEC customers. (c) Using a t or percentile interval, we would tend to understate the difference in mean repair times and perhaps fail to recognize that the mean repair times for Verizon customers are significantly shorter than for CLEC customers.

18.39 (a) A traditional 95% one-sample t confidence interval is $(59.83, 66.19)$. (b) The value 92.3 is an outlier and might strongly influence the traditional confidence interval. (c) The 95% percentile interval is $(60.43, 66.32)$. Both ends are to the right of the interval in (a) due to the skewness in the sampling distribution. This is also slightly narrower. (d) A 95% confidence interval for the mean weights of male runners (in kilograms) is $(60.43, 66.32)$.

18.41 (a) Two large outliers are present. The t procedures can be used even for clearly skewed distributions when the sample size is large, roughly $n \geq 40$. In this example, $n = 43$, so one-sample t procedures may be safe. (b) A traditional 95% one-sample t confidence interval is $(116.3, 124.8)$.

(c) The bootstrap distribution shows moderate skewness to the right; a bootstrap t interval should be moderately accurate. (d) The 95% percentile interval is (116.9, 124.4). This agrees closely with the interval found in (b), so we conclude that the one-sample t interval is reasonably accurate here.

18.43 (a) The data are clearly right-skewed. The mean would not be a useful measure of the price of a typical house in Ames. The trimmed mean or the median might be more useful. We examine the trimmed mean. (b) The standard error of our bootstrap statistic is 6743. (c) The distribution looks approximately Normal, and the bias is relatively small. The 95% bootstrap t interval and the 95% bootstrap percentile interval are reasonable choices. The 95% percentile interval is (119,677, 144,161). For comparison, the 95% BCa interval is (120,484, 147,562), the 95% tilting interval is (120,496, 146,665), and the 95% bootstrap t interval is (119,709, 146,753). (d) We are 95% confident that the mean selling price of all homes sold in Ames for the period represented by these data is in the interval ($119,677, $144,161).

18.45 (a) The relationship appears linear and the association is negative: $r = -0848$. (b) The 95% BCa interval is (−0.898, −0.769), and the 95% tilting interval is (−0.900, −0.776). All should be accurate intervals. They provide a 95% confidence interval for the population correlation between weight and gas mileage in miles per gallon for all 1990 model year cars. (c) The least-squares regression line to predict gas mileage from weight is mileage = 48.35 − (0.0082)(weight). The traditional 95% t confidence interval for the slope is (−0.0096, −0.0068). (d) The bootstrap interval is (−0.00954, −0.00676).

18.47 (a) Examining the plots, we see that the bootstrap distribution with the outlier included is shifted significantly to the left (centered at a smaller value) of the bootstrap distribution with the outlier excluded. Also, the bootstrap distribution with the outlier included appears to be slightly left-skewed. There is little bias in either case (in fact, any apparent bias is due to random resampling, because the sample mean has no true bias). (b) A 95% BCa interval for the mean with the outlier included is (16,504, 18,322). A 95% BCa interval for the mean with the outlier removed is (17,241, 18,614). The lower confidence limit for the interval based on the data that include the outlier is much smaller than the lower confidence limit for the interval based on the data with the outlier excluded. The outlier was an unusually small value, so it appears that the effect of the outlier is to pull the lower limit down. The upper confidence limits for both intervals are more nearly equal, but the upper confidence limit for the interval based on the data that include the outlier is smaller than the upper confidence limit for the interval based on the data with the outlier excluded. Thus, the small value of the outlier also pulls the upper confidence limit down.

18.49 (a) $H_0: \mu_1 = \mu_2$, $H_a: \mu_1 \neq \mu_2$. (b) The P-value is 0.423. (c) The P-value is 0.444. This is consistent with the P-value we computed in (b). We conclude that there is little evidence that the population means μ_1 and μ_2 differ. (d) A 95% BCa confidence interval for the change from 2001 to 2002 is (−39.3, 156.7). This interval includes 0 and suggests that the two means are not significantly different at the 0.05 level. This is consistent with the conclusions in (c).

18.51 The standard deviation for the estimated P-value of 0.015 for the DRP study based on $B = 999$ resamples is 0.00385. The standard deviation for the estimated P-value of 0.0183 based on the 500,000 resamples in the Verizon study is 0.000190.

18.53 (a) H_0: $p_1 = p_2$, H_a: $p_1 > p_2$. (b) The z statistic is $z = 2.43$. The P-value for the test is 0.0075. (c) Under the null hypothesis, all 170 firms are equally likely to be successful. That is, successes occur for reasons that have nothing to do with whether the firm has an exclusive-territory clause. We can resample in a way consistent with the null hypothesis by choosing an ordinary SRS of 142 of the firms without replacement and assigning them to the exclusive-territory clause group. The P-value for the permutation test is 0.018. This P-value is larger than that found in part (b), largely because the z test fails to take ties into account. (d) There is evidence at the 0.05 level that exclusive-territory clauses increase the chance of success. There is not evidence that exclusive-territory clauses increase the chance of success at the 0.01 level. (e) A 95% BCa confidence interval for the difference between the two population proportions is (0.033, 0.426). This interval does not include 0 and lies to the positive side of 0. This is consistent with the results of the permutation test in part (d), which rejected the null hypothesis at the 0.05 level.

18.55 (a) Let ρ denote the correlation between the salaries and batting averages of all Major League Baseball players. We test the hypotheses H_0: $\rho = 0$, H_a: $\rho > 0$. (b) The P-value is 0.257. We conclude that there is not strong evidence that salaries and batting averages are correlated in the population of all Major League Baseball players.

18.57 (a) For the median we test the hypotheses H_0: median time for right hand = median time for left hand, H_a: median time for right hand $\neq$ median time for left hand. For the 25% trimmed mean we test the hypotheses H_0: 25% trimmed mean time for right hand = 25% trimmed mean time for left hand, H_a: 25% trimmed mean time for right hand $\neq$ 25% trimmed mean time for left hand. (b) The permutation distribution is clearly not Normal. The P-value for the permutation test for the difference in medians is 0.002. (c) The permutation distribution looks much more like a Normal distribution than the permutation distribution in (b) for the difference in medians. For the permutation test for the difference in 25% trimmed means, P-value = 0.002. (d) There is strong evidence that there is a difference in the population median times when using the right hand versus when using the left hand. There is strong evidence that there is a difference in the population 25% trimmed mean times when using the right hand versus when using the left hand.

18.59 (a) Let p_1 denote the proportion of women in the population who pay attention to a "No Sweat" label when buying a garment and p_2 denote the proportion of men in the population who pay attention to a "No Sweat" label when buying a garment. We test the hypotheses H_0: $p_1 = p_2$, H_a: $p_1 \neq p_2$. (b) The P-value for the permutation test is 0.002. (c) The permutation distribution is approximately Normal (except that it is discrete; you can see this using a Normal quantile plot or by observing spikes in the histogram).

Solutions to Odd-Numbered Exercises

18.61 A two-sided permutation test of the hypothesis has P-value = 0.004, and we conclude that there is strong evidence that there is a correlation between square footage and age of a house in Ames, Iowa.

18.63 A 95% z interval using the observed difference in proportions (0.2364) and the bootstrap standard error (SE = 0.215) is (−0.185, 0.657).

18.65 (a) We performed a two-sided permutation test on the ratio of standard deviations. Some of the ratios were infinite because the permutation test produced a standard deviation of 0 in the denominator. The mean and SE were not given, but we were still able to give P-value = 0.386 based on the permutation distribution. This P-value tells us that there is not strong evidence that the variability in the repair times for ILEC and CLEC customers differ. (b) The P-value for the permutation test differs from that obtained by the F statistic. This suggests that the test based on the F statistic is not very accurate.

18.67 (a) We test the hypotheses $H_0: \mu = 0$, $H_a: \mu < 0$. Note that negative values of μ indicate that mean posttest scores are higher than mean pretest scores, and hence that test scores have improved. (b) The P-value is 0.036, so there is evidence (significant at the 0.05 level but not at the 0.01 level) that the mean change (pretest−posttest) is negative and hence that posttest scores are higher, on average, than pretest scores. (c) The area to the left of −1.45 in your graph is the P-value.

18.69 (a) For a Uniform distribution on 0 to 1, the population median is 0.5. The bootstrap distribution appears to be bimodal, not Normal (you may get a different picture, depending on the random data you generate). (b) The bootstrap standard error is 0.072. A 95% bootstrap t confidence interval is (0.373, 0.572). (c) The bootstrap 95% BCa confidence interval is (0.369, 0.620). The 95% bootstrap t confidence interval is somewhat wider than the 95% BCa interval. This, and the fact that the bootstrap distribution for the median is not Normal, show that the bootstrap t interval is not reliable here.

18.71 (a) The bootstrap distribution is left-skewed and does not appear to be approximately Normal. The bootstrap standard error is 0.139. (b) The bootstrap t confidence interval is not appropriate here because the bootstrap distribution is not approximately Normal. (c) A 95% BCa confidence interval is (0.179, 0.710).

18.73 (a) The histogram of the 2000 data is strongly right-skewed with two outliers, one of which is extreme. This violates the guideline for using the t procedures given in Section 17.1; namely, *for a sample size of at least 15, t procedures can be used except in the presence of outliers or strong skewness*. The histogram of the 2001 data is right-skewed, but less strongly than that of the 2000 data. (b) The P-value for the permutation test for the difference in means is 0.302. We conclude that there is not strong evidence that the mean selling prices for all Seattle real estate in 2000 and in 2001 are different.

18.75 The study described in Exercise 18.74 is a one-sample problem. We have no methods for carrying out a permutation test in such one-sample problems (there is no obvious way to resample that is consistent with a one-sample test for a mean).

18.77 A 95% bootstrap t confidence interval for the mean change (After − Before) is (0.412, 0.865). Zero is outside this interval, so the result is significant at the 0.05 level. We conclude that there is strong evidence that the mean difference is different from 0.

18.79 (a) The proportion of girls who like chocolate ice cream is 0.80. The proportion of boys who like chocolate ice cream is 0.667. (b) The P-value for a two-sided permutation test is 0.222. There is not strong evidence that there is a difference in the proportion of boys and girls who like chocolate ice cream.

18.81 We test the hypotheses H_0: $\mu_1 = \mu_2$, H_a: $\mu_1 > \mu_2$. The P-value is 0.209. Thus, there is not strong evidence that the mean word count is higher for ads placed in magazines aimed at people with high education levels than for ads placed in magazines aimed at people with medium education levels. The 95% confidence interval in Exercise 18.80(d) for the difference in means contained 0. This suggests that there is not strong evidence of a *difference* in mean word counts. Here we conclude that there is not strong evidence that the mean word count is *higher* for ads placed in magazines aimed at people with high education levels than for ads placed in magazines aimed at people with medium education levels.

18.83 (a) The bootstrap distribution appears to be approximately Normal. (b) The bootstrap standard error is 4.603. A 95% bootstrap t confidence interval using the conservative method for degrees of freedom is (−13.43, 6.09). Using $n_1 + n_2 - 2$ degrees of freedom, the interval becomes (−12.89, 5.55). (c) A 95% percentile interval is (−12.48, 5.56). This agrees closely with the second interval found in (b), so we conclude that the intervals are reasonably accurate. These intervals include 0, and so we would conclude that there is not strong evidence (at the 0.05 level) of a *difference* in the mean monthly burglary counts. The tests in Exercise 18.82 were one-sided tests and showed no evidence of a *decrease* in mean monthly burglaries (or of an *increase* in the case of part (d) of Exercise 18.82).